KT-484-313

Water Regulations in Brief

Ray Tricker

NORWICH CITY COLLEGE		
Stock No.	240128	
Class	696.1 TRI	
Cat.	A2	Proc	3WL

ELSEVIER

AMSTERDAM • BOSTON • HEIDELBERG • LONDON • NEW YORK • OXFORD
PARIS • SAN DIEGO SAN FRANCISCO • SINGAPORE • SYDNEY • TOKYO

Butterworth-Hienemann is an imprint of Elsevier

Butterworth-Heinemann is an imprint of Elsevier
The Boulevard, Langford Lane, Kidlington, Oxford OX5 1GB, UK
30 Corporate Drive, Suite 400, Burlington, MA 01803, USA

First edition 2009

Copyright © 2009, Ray Tricker. Published by Elsevier Ltd. All rights reserved

The right of Author Name to be identified as the author of this work has been
asserted in accordance with the Copyright, Designs and Patents Act 1988

No part of this publication may be reproduced, stored in a retrieval system
or transmitted in any form or by any means electronic, mechanical, photocopying,
recording or otherwise without the prior written permission of the publisher

Permissions may be sought directly from Elsevier's Science & Technology Rights
Department in Oxford, UK: phone (+44) (0) 1865 843830; fax (+44) (0) 1865 853333;
email: permissions@elsevier.com. Alternatively you can submit your request online by
visiting the Elsevier web site at http://elsevier.com/locate/permissions, and selecting
Obtaining permission to use Elsevier material

Notice
No responsibility is assumed by the publisher for any injury and/or damage to persons
or property as a matter of products liability, negligence or otherwise, or from any use
or operation of any methods, products, instructions or ideas contained in the material
herein. Because of rapid advances in the medical sciences, in particular, independent
verification of diagnoses and drug dosages should be made

British Library Cataloguing in Publication Data
A catalogue record for this book is available from the British Library

Library of Congress Cataloging-in-Publication Data
A catalog record for this book is available from the Library of Congress

ISBN: 978-1-85617-628-6

For information on all Butterworth-Heinemann
publications visit our web site at books.elsevier.com

Typeset by Macmillan Publishing Solutions
www.macmillansolutions.com

Printed and bound in Italy

09 10   10 9 8 7 6 5 4 3 2 1

Working together to grow
libraries in developing countries

www.elsevier.com | www.bookaid.org | www.sabre.org

ELSEVIER   BOOK AID
International   Sabre Foundation

Contents

Water Regulations in Brief

Please return on or before the last
date shown below

240   128

Preface

Many books have been written about the UK's Building Act, Building Regulations and their associated Approved Documents and Bye-laws. In a similar manner, there have been a number of attempts to describe the requirements of the Water Supply (Water Fittings) Regulations 1999 and the Water Bye-laws 2000 (Scotland), the most famous, most widely read and considered the best of which is the Water Regulations Advisory Service's *Water Regulations Guide*.

None of these books, however, actually links the current and latest editions of the Building Regulations, Water Regulations, Water Bye-laws and Wiring Regulations together in one book. The aim, therefore, of *Water Regulations in Brief* is to provide an all-encompassing reference to the requirements of the water industry and for water fitting designers, manufacturers, suppliers and installers of water systems in accordance with the Building Act.

1 What is the aim of the Water Regulations and the Water Bye-laws (Scotland)?

The Water Regulations/Bye-laws set requirements for the design, installation and maintenance of water fittings and plumbing systems which are enforced by water companies in their respective areas of supply. The water companies have a legal obligation to ensure that these Regulations/Bye-laws have been implemented, including the inspection of new and existing installations.

The Regulations/Bye-laws apply to any water fitting installed or used (or to be installed and/or used) in premises to which water is (or is to be) supplied by a water undertaker. They are aimed at preventing waste, undue consumption, misuse and/or contamination of water supplied by a water undertaker.

These Regulations/Bye-laws do *not*, however, apply to:

- water fittings which are not connected to water supplied by a water undertaker;

- a water fitting installed or used in connection with water supplied for purposes other than domestic or food production purposes, provided that:
  - the water is metered;
  - the supply of the water is for a period not exceeding one month; and
  - no water can return through the meter to any pipe that is the responsibility of the water undertaker;
- certain water fittings in connection with water supplied for non-domestic
or food production purposes, neither do they require any person to
remove, replace, alter, disconnect or cease to use any water fitting which
was lawfully installed or used, or capable of being used, before the current
Regulations and Bye-laws came into force.

2 How do the Water Regulations affect you?

Water systems and fittings in premises that are, or will be, connected to the
public water supply must comply with the Regulations/Bye-laws. If you are
planning to carry out certain plumbing work, you must obtain the prior consent of your water company by giving advanced notice of the work. This
includes the installation of water fittings in connection with the:

- construction of a large pond or swimming pool with automatic replenishment;
- erection of any new building or structure;
- extension or alteration of the water system in any premises except a
domestic dwelling;
- material change in use of any premises;
- installation of any fitting listed in section 5 of the Regulations.

3 Building legislation

The prime purpose of building legislation is to prevent waste, undue consumption, misuse and contamination of water, and to assist in the conservation of fuel and power. It imposes on owners and occupiers of buildings a
set of requirements concerning the design and construction of buildings and
the provision of services, fixtures and fittings used in (or in connection with)
buildings. These requirements involve, and cover:

- a method of controlling (inspecting and reporting) buildings and their
services, fittings and equipment;
- how services, fittings and equipment may be used;
- the inspection and maintenance of services, fittings or equipment used.

4 What is the purpose of the Building Regulations?

The Building Regulations are legal requirements laid down by parliament, based on the Building Act 1984. They are approved by parliament and deal with the minimum standards of design and building work for the construction of domestic, commercial and industrial buildings.

The Building Regulations ensure that new developments or alterations and/or extensions to buildings are all carried out to an agreed standard that protects the health and safety of people in and around the building.

From the point of view of water systems, Building Regulations primarily include water services, fixtures, fittings and equipment, but also cover:

- cesspools (and other methods for treating and disposing of foul matter);
- drainage (including waste disposal units);
- materials and components (suitability, durability and use);
- prevention of infestation;
- resistance to moisture and decay;
- waste (storage, treatment and removal);
- wells and boreholes for supplying water;
- electrical safety;
- emission of gases, fumes, or other noxious and/or offensive substances;
- fire precautions

and matters connected with (or ancillary to) any of the above matters.

Builders and developers are required by law to obtain Building Control approval, which is an independent check that the Building Regulations have been complied with. There are two types of Building Control providers: the local authority and approved inspectors.

5 How do the Building Regulations affect water fittings and water supplies?

The Building Act stipulates that plans for proposed buildings will ensure that all occupants of the house shall be provided with a supply of 'wholesome water, sufficient for their domestic purposes'. This can be achieved by:

- connecting the house to water supplies from the local water authority (normally referred to as the 'statutory water undertaker');
- otherwise taking water into the house by means of a pipe (e.g. from a local recognized supply);
- providing a supply of water within a reasonable distance from the house (e.g. from a well).

If an occupied house is not within a reasonable distance of a supply of 'wholesome water' or if the local authority is not satisfied that the water supply is capable of supplying 'wholesome water', then they can give notice that the owner of the building must provide water within a specified time. They also have the authority to prohibit the building from being occupied.

6 What about plumbing?

Although planning permission is not required for plumbing replacements, it would always be wise to consult the technical services department for any installation that alters present internal or external drainage. Building Regulation approval is, however, required for the installation or replacement of any hot water system if the water heater is unvented (i.e. supplied directly from the mains without an open expansion tank and with no vent pipe to the atmosphere) and has a storage capacity greater than 15 litres.

Building standards are enforced by the local Building Control officer, but for matters concerning drainage or sanitary installations, you will need to consult their technical services department.

7 What are Approved Documents and Technical Handbooks?

These are a series of documents that are intended to provide practical guidance with respect to the requirements of the Building Regulations, which impose on owners and occupiers of buildings a set of requirements concerning the design and construction of buildings and the provision of services, fittings and equipment used in (or in connection with) buildings. These involve, and cover:

- a method of controlling (inspecting and reporting) buildings;
- how services, fittings and equipment may be used;
- the inspection and maintenance of any service, fitting or equipment used.

Details of these requirements are available as a series of documents called Approved Documents and Technical Handbooks.

8 How do the requirements of the Building Regulations affect water services?

The Building Regulations cover all new building work. This means that if you want to provide new and/or additional fittings in a new building, extension or alterations such as drains or heat-producing appliances, washing and sanitary facilities and hot water storage (particularly unvented hot water systems), the Building Regulations will probably apply. The mandatory requirements of the Regulations concerning water are:

- site preparation;
- resistance to moisture;
- protection against sound within a dwelling-house, etc.;
- sanitary conveniences and washing facilities;
- bathrooms;
- drainage and waste disposal;
- rainwater drainage;
- foul water drainage;
- separate systems of drainage;
- wastewater treatment systems and cesspools;
- building over sewers;
- hot water storage;
- conservation of fuel and power;
- performance tests;
- electrical earthing.

Even when planning permission is not required, most building works, including alterations to existing structures, are subject to minimum standards of construction to safeguard public health and safety.

9 Structure of this book

The content of the book is split into two parts together with a list of relevant legislation, abbreviation list and index, as follows.

Part I

Part I is intended as a background reminder to where water comes from, and how it is collected, treated and used. It also shows how the Building Regulations have a great effect on the design, installation and servicing of water fittings. For this purpose, Part I has been split into three chapters, as follows:

| Chapter 1 | Introduction | Background information concerning the sources of water, potable water, water pollution, duties of the water companies and the Water Regulations Advisory Scheme |

| Chapter 2 | What are the Water Regulations? | A brief description of the regulations and bye-laws concerning water supplies in the UK, together with details of the mandatory requirements for water fittings |
| Chapter 3 | What are the Building Regulations? | Information concerning building legislation, the purpose of the Building Regulations and how they affect, interact with and relate to water supplies, services and the Water Regulations |

Part II

Part II is the main part of the book, covering all of the relevant regulations and requirements for water fittings and water services that are contained in the UK's most recent Water, Building and Wiring Regulations. As this book is primarily intended as a reference book, explanatory text has been kept to a minimum and regulatory requirements have been tabulated for easy reference.

Part II is intended to be an all-encompassing reference to anything involving the source, supply, use and disposal of water, and is aimed at designers, installers and those responsible for the operation and maintenance of water fittings. Similar to my other 'In Brief' books (and following requests and suggestions from previous readers) this part of the book's layout has been structured around the design of a building (i.e. starting from the foundations and working upwards). This has been deliberately done so as to enable the reader (whether they be an architect, a professional water engineer, a do-it-yourselfer or a student) to quickly locate a particular part of a building (e.g. a bathroom) and to concentrate on those requirements and regulations that are applicable to that specific area or topic.

For convenience, therefore, Part II consists of 12 chapters as follows:

| Chapter 4 | Meeting the requirements of the Building Regulations | Background information concerning materials and workmanship, technical specifications, standards, technical approvals and independent certification schemes |
| Chapter 5 | Design and installation | A description and listing of all of the mandatory design requirements for water fittings and water supplies and the commissioning and testing of installations |

Chapter 6	Site preparation	All of the requirements and regulations concerning subsoil drainage, groundwater, surface water, existing drains and water fittings that are laid below ground level
Chapter 7	Drainage	By its very nature, this is an extremely large chapter that covers all the rules about drainage including the requirements for surface water drainage, rainwater drainage, wastewater drainage, foul water drainage, wastewater treatment systems, cesspools, existing sewers, septic and holding tanks, private wastewater treatment plants and septic tanks
Chapter 8	Ventilation	A short chapter describing the requirements for, and the types of, ventilation systems required to limit the accumulation of moisture from a building, controlling excess humidity, dispersing residual water vapour and extracting water vapour from wet areas (e.g. bathrooms)
Chapter 9	Floors	A brief chapter concerning the requirements for floors being capable of resisting ground moisture and being moisture resistant. It includes the requirements (especially in Scotland) for wastewater lifting plants (in sanitary appliances below flood level), washing down areas and hot water underfloor heating
Chapter 10	Roofs	This chapter covers the requirements to guard the building against the effects of precipitation and the need for surface water drainage
Chapter 11	Chimneys and fireplaces	A few lines on the need to guard against moisture entering the building via a chimney

Chapter 12	Sanitary facilities	I suspect that this will probably be the most used chapter of the whole book (especially by plumbers) as it contains all of the Water, Building and Wiring Regulations concerning sanitary facilities, water closets, urinals, bathrooms and shower rooms, as well as the need to provide adequate access to and use of these facilities by disabled people
Chapter 13	Hot water storage systems	A chapter covering the requirements for heating and hot water systems, insulation of pipes, energy efficiency, etc.
Chapter 14	Cold water storage systems	Mandatory requirements extracted from the Water Regulations concerning the storage of cold water
Chapter 15	Heating systems	The final chapter concerns the requirements for heating and hot water systems, boilers, system efficiency and commissioning services (metering and fuel storage, etc.)

The two parts of this book are supported by the following annexes:

- Annex A: Acronyms and abbreviations;
- Annex B: Legislation and standards – containing details of all the relevant EU Harmonized Directives, UK Legislation, British and European Standards affecting the Water Regulations/Bye-laws;

plus a full index.

It is hoped that the following symbols will help you to get the most out of this book:

Need to be careful (e.g. a necessary requirement, potential minefield or legal/statutory requirement).

A good idea or a useful reminder.

Note: Additional advice and guidance.

I have also highlighted the actual requirements extracted from the Water, Building and Wiring Regulations as per the example below:

> Any water fitting laid below ground level shall have a depth of cover sufficient to prevent water freezing in the fitting.   SI 1148 7(4)

For your convenience (and to save your constantly having to flick backwards and forwards through the book to find the correct requirement), many of these requirements have been shown more than once (i.e. in different chapters of the book), as have a few of the figures and tables.

 Note: If any reader has any thoughts about the contents of this book (such as areas where perhaps they feel I have not given sufficient coverage, or there are omissions, mistakes, etc.) then please let me know by e-mailing me at ray@herne.org.uk and I will make suitable amendments in the next edition of this book.

Thank you and I hope that you find this unique reference book useful in your work.

Ray

About the Author

Ray Tricker MSc, IEng, CMan, FIET, FCMI, FCQI, FIRSE is the Principal Consultant of Herne European Consultancy Ltd, a company specializing in quality, environmental and safety management systems. He is also an established Butterworth-Heinemann author (27 titles). He served with the Royal Corps of Signals for a total of 37 years, during which time he held various managerial posts culminating in being appointed as the Chief Engineer of NATO's Communication Security Agency (ACE COMSEC).

Most of Ray's work since joining Herne has centred on the European Railways. He has held a number of posts with the Union International des Chemins de Fer (UIC) [e.g. Quality Manager of the European Train Control System (ETCS)] and with the European Union (EU) Commission [e.g. T500 Review Team Leader, European Rail Traffic Management System (ERTMS) Users' Group Project Co-ordinator, HEROE Project Coordinator]. Currently, as well as writing books on diverse subjects as Optoelectronics, Medical Devices, ISO 9001:2000, Building, Wiring and Water Regulations for Elsevier under their Butterworth-Heinemann and Newnes imprints, he is busy helping small businesses from around the world (usually on a no-cost basis) to produce their own auditable quality and/or integrated management systems to meet the requirements of ISO 9001:2008, ISO 14001 and OHSAS 18001, etc. He is also a UKAS Assessor for the assessment of certification bodies for the harmonization of the Trans-European, High Speed Railway Network.

Recently he has been appointed as the Quality, Safety and Environmental Manager for the Project Management Consultancy overseeing the multibillion-dollar Trinidad Rapid Rail System. One day he may retire!

To my grandson, James

Part I

1

Introduction

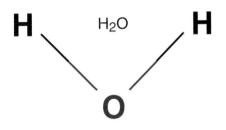

Figure 1.1 Water (molecule)

Water (or H_2O from a chemical point of view; Figure 1.1) is a simple compound made up of two hydrogen atoms attached to a single oxygen atom. The composition of water was discovered by a London scientist named Henry Cavendish in 1776. It is normally transparent, odourless, tasteless and freely available. But although water and ice may appear to be colourless in small quantities, they actually have a very light blue hue.

Behind hydrogen (H_2), water is the second most common molecule in the universe and no living organism can exist without it; indeed, as much as half the weight of plants and animals is made up of water. Water is in a constantly moving cycle (known scientifically as the hydrological cycle; Figures 1.2 and 1.3) which consists of a continuous exchange of water within the hydrosphere, and between the atmosphere, surface water, groundwater, soil water and plants. Water moves perpetually through each of these regions and the cycle consists of the following transfer processes:

- evaporation from oceans and other water bodies into the air;
- transpiration from land plants and animals into the air;
- precipitation from water vapour condensing from the air and falling back to earth or onto the ocean.

As the sun beats down, some of the surface water evaporates; this water vapour rises as part of the air and is moved along by the wind. Should it pass

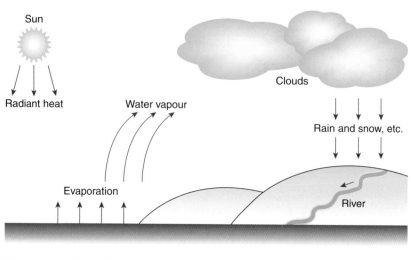

Figure 1.2 Simplified water cycle

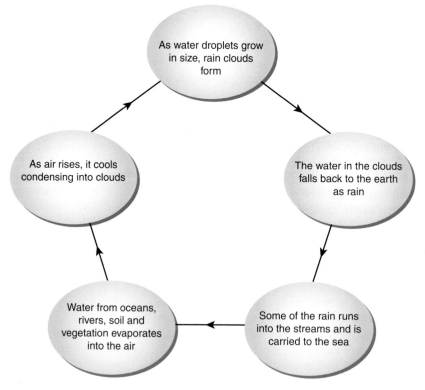

Figure 1.3 The hydrological cycle

over a land mass it may become a cloud and as more moisture is attracted to the cloud, or the cloud passes over rising ground, the water particles become larger and fall as rain, sleet or snow.

1.1 Rain

As the rain comes into contact with the ground there are several avenues open to it. It may be revaporized and return to the atmosphere, be absorbed by the ground, or remain on the surface of the ground and run downhill, forming streams and rivers which will run into the sea, and the cycle begins once more (Figure 1.4).

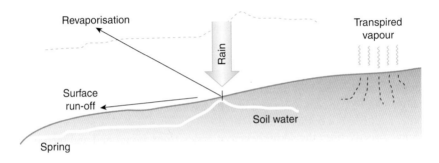

Figure 1.4 The effect of rain

Normally, streams and rivers present a placid picture, but when there is too much water in the stream or river, or when there is a very heavy down-fall, they can flood and (as been experienced in recent years) cause enormous damage, not just through the amount of water but also from the amount of soil carried by the water from fields and other rubbish picked up on its way. When it rains particularly hard, the drains and soil are unable to soak up water quickly enough and local flooding will result.

1.2 How will climate change affect water?

Climate change, with the probability of longer periods of weather events such as storms and droughts, will have a huge impact on water. Storms and intense rainfall will cause surface flooding rather than long-term infiltration to underground aquifers, and this will carry pollutants from the land and deposit them into rivers, lakes, etc. Droughts can create a water shortage that will influence the amount of treated wastewater that can be discharged into rivers as effluent, which will result in a lower flow rate and less dilution and thus reduce the quality of the diluted water.

As water is directly dependent on the natural environment, everyone needs to be aware of the extent of the problem of climate change. There needs to be a concerted effort to reduce the amount of energy consumption used to obtain the water and thus (i.e. through carbon management) a reduction in the contribution to climate change caused by 'farming water'.

1.3 What about drinking water?

The human body can contain anything between 55 and 78 per cent of water, depending on body size. To function properly, the body requires topping up each day to avoid becoming dehydrated and running the risk of headaches, tiredness, loss of concentration, constipation and the possibility of suffering from kidney stones.

The British Dietetic Association has stated that the average adult should consume a minimum of 4.5 pints (2.5 litres) of liquid per day and that this should, ideally, be plain drinking water (most carbonated or still drinks are only about 65 per cent water and usually have a high percentage of sugar and other additives) or bottled water, which can be either spring water (water collected directly from the spring where it arises from the ground and then bottled at the source) or mineral water (which, having emerged from the ground, flows over rocks before it is collected, resulting in a higher content of various minerals.)

There has been and still is a lot of controversy about the benefits of drinking bottled water instead of water straight from the tap. Provided that the current standards set out in the Water Supply Regulations are observed, there is nothing to suggest that, in the UK, there will be anything wrong with tap water. Indeed, according to a recent issue of *Which? Online*, 98.8 per cent of UK tap water sampled passed Drinking Water Inspectorate tests and there were no obvious health benefits of bottled water over tap water.

The Drinking Water Directive (DWD 98/83/EC) concerns the quality of water intended for human consumption and aims to protect the health of consumers. The DWD identifies 48 microbiological and chemical parameters that must be monitored and tested regularly to ensure that water is wholesome, healthy, clean and tasty throughout the European Union (EU).

Member states of the EU may include additional requirements that are relevant within their own territory (such as the use of chlorine and/or fluoride) and they are entitled to set higher standards. They must, however, meet the minimum levels of the Directive, particularly in respect of bacteriological contamination, chemical substances (including construction products) in contact with drinking water and small water supplies. They must also pay due attention to risk assessment and risk management.

Note: The requirements for water quality (i.e. as extracted from the Water Regulations) are as follows:

> All premises supplied with water for domestic purposes   SI 1148-26
> shall have at least one tap conveniently situated for the
> drawing of drinking water.
> A drinking water tap shall be supplied with water from:
> * a supply pipe;   SI 1148-27(a)
> * a pump delivery pipe drawing water from a supply pipe; or   SI 1148-27(b)
> * a distributing pipe drawing water exclusively from a   SI 1148-27(c)
>   storage cistern supplying wholesome water.

1.4 Where is the Earth's freshwater located?

Water exists in the air as water vapour and in the ground in underground lakes and wells (commonly referred to as aquifers), which are recharged by precipitation and ground seepage. In all, about 70 per cent of the Earth's surface is covered in water but nearly 98 per cent of this is saline water in the oceans, while the remaining 2 per cent of freshwater provides people (and other life forms) with most of the water they need every day to live.

The total freshwater supply for the world is in the region of 300 trillion gallons, the majority of which is stored on the ground where it is available in reservoirs, streams, rivers, lakes, etc. A further 3000 trillion gallons is also available in the form of atmospheric water vapour which will eventually fall as precipitation. Conversely, about 250 trillion gallons of water evaporates into the atmosphere every day (see Figures 1.2 and 1.3).

Table 1.1 provides a detailed look at where the Earth's water is located as a percentage of total water. Even though rivers are a very minor source, they are nevertheless the most used source worldwide.

1.5 Potable water

Drinking water is becoming a rare and sought-after commodity and reducing waste by using drinking water only for human consumption is fast becoming the preferred option. Indeed, in some cities, such as Hong Kong, seawater is extensively used for flushing toilets to conserve freshwater resources.

UK tap water is recognized as one of the best in the world, but we need to look after it in our homes and this is best done by:

* keeping taps clean, to stop the water becoming contaminated;

- making sure that kitchen appliances do not damage the taste of drinking water;
- making sure that when plumbing is changed, altered or modified only fittings approved for use with drinking water are used.

In other words, complying with the requirements of the Water and Building Regulations!

Table 1.1 **Water sources**

Water source	Water volume (cubic miles)	Per cent of total water	Per cent of freshwater
Oceans, seas and bays	321,000,000	96.5	
Ice caps, glaciers and permanent snow	5,773,000	1.74	68.7
Groundwater	5,614,000	1.7	
Fresh	2,526,000	0.76	30.1
Saline	3,088,000	0.94	
Soil moisture	3,959	0.001	0.05
Ground ice and permafrost	71,970	0.022	0.86
Lakes	42,320	0.013	
Fresh	21,830	0.007	0.26
Saline	20,490	0.006	
Atmosphere	3,095	0.001	0.04
Swamp water	2,752	0.0008	0.03
Rivers	509	0.0002	0.006
Biological water	269	0.0001	0.003

Water fit for human consumption is called drinking water or potable water. Water that is not potable can be made potable by filtration or distillation (heating it until it becomes water vapour, and then capturing the vapour without any of the impurities it leaves behind), or by other methods (such as chemical and/or heat treatment) to kill the bacteria. About one-third of the UK's drinking water is extracted from groundwater, which also feeds the springs from which streams and rivers flow.

The European Water Framework Directive (2006/118/EC) is aimed at protecting groundwater and has recently been improved to include proposals aimed at reducing the contamination of surface water (rivers, lakes, estuaries and coastal waters) by pollutants. The Directive also includes an objective to meet 'good chemical status' by 2015 and to prevent, or at least limit, the input of pollutants to groundwater.

1.6 What about contamination of drinking water?

As a natural resource, water is becoming increasingly scarce in certain places and its availability is a major social and economic concern. On the other side of the coin, in the developing world, 90 per cent of all wastewater still goes untreated into local rivers and streams and, currently, about one billion people around the world routinely drink unhealthy water.

1.6.1 Water pollution

Normally, when you think about river pollution, you assume that it comes from places such as factories, farms and industry. However, a large proportion of all river pollution currently comes from people's homes because of incorrect plumbing which frequently results in wastewater from dishwashers, washing machines, sinks, baths and even toilets being flushed directly into a local river. These 'misconnected' pipes are a common cause of pollution to rivers and streams, especially in towns and cities. Pollution may also be the result of discharges through pipes, run-off from streets and buildings, or agricultural nutrients lost from fields.

Many human activities and their by-products have the potential to pollute water, as do large and small industrial enterprises (even the water industry), and pollution results from deliberate or accidental incidents in agriculture, horticulture, transport or the urban infrastructure, as well as discharges from abandoned mines and other underground contaminated sources.

1.6.2 Drainage

There are normally two forms of drainage: surface water and foul water.

- **Surface water drains** or 'storm drains' carry rainwater from road surfaces and rooftops into local rivers and streams that then flows into the river untreated.
- **Foul water drains** carry wastewater from toilets, sinks, baths and household appliances to the sewage treatment works. This water is then treated before it can safely flow back into river and streams.

In some houses, however (particularly those built during the nineteenth century), there is often a combined drainage system where foul and surface water all drain into the foul sewer, which is then connected to a sewage treatment plant. If there is a fracture or a blockage in one of these connecting pipes, untreated foul water can escape and contaminate groundwater.

Occasionally (i.e. through bad plumbing) household appliances are accidentally connected to a surface water drain instead of the foul water drain, and wastewater from sinks, toilets and washing machines goes straight into watercourses.

 Untreated sewage effluent in the water causes oxygen levels to drop drastically. This will cause sewage fungus to cover the bed of the watercourse like a blanket, so that the river can no longer support fish, insects and other animals that live in and around the water.

1.6.3 Sewerage and storm water treatment works

Although more than £15 billion has apparently been spent by the water companies in the past 18 years in trying to improve the dilapidated sewage and storm water treatment works that existed upon privatization in 1990, 80 per cent of British rivers still fail to meet European standards.

1.6.4 Rivers

Over one-third of the UK's rivers still need improvement and one in six rivers in urban areas are still classed as 'poor' or 'bad'.

1.7 What about the water industry?

The water industry provides drinking water and wastewater services (including sewage treatment) to households and industry. Drinking water is mainly collected at springs, or extracted from artificial borings in the ground or from wells. Other water sources such as rainwater, streams, rivers and lake water are available but all surface water will need to be purified for human consumption to remove any undissolved substances and harmful microbes, etc., by filtering, chlorination, boiling and/or distillation.

One of the problems that the water authorities have to contend with is that the water industry is extremely energy intensive and consumes about 3 per cent of the total energy used in the UK. Most of this is used to pump water and wastewater and to run treatment plants that ensure water meets strict environmental and health quality standards. The water industry is also responsible for approximately 4 million tonnes of greenhouse gas emissions (CO_2 equivalent) every year and although this is less than 1 per cent of total UK emissions it is nevertheless increasing every year owing to population and consumption growth.

On the plus side, however, rather than relying on the heavy hand of government law to reduce carbon waste, the water industry is trying to find other ways around the problem by a series of energy efficiency projects such as increasing the recycling of biosolids (i.e. sewage sludge) and using renewable sources of energy such as wind power, combined heat and power (CHP) and biofuels (although it has been said that even though biofuels may be thought of as energy efficient, the production costs far exceed their benefits).

Water UK (which represents all of the UK water and wastewater service providers at national and European level) agrees that surface water management plans will become extremely important in the future and supports the argument for greater use of sustainable urban drainage systems (SUDS) to provide surface water treatment and thus delay the flow into drains.

1.8 What problems do the water companies have to contend with?

Although water is obviously the water companies' principal commodity, as well as being a friend it can be a very serious and costly enemy, and is a major cause of failure in water supply equipment and facilities such as water wells, rainwater harvesting cisterns, water supply networks, water purification facilities, water tanks, water towers and water pipes, as well as some old aqueducts. Associated electrical, electronic and electromechanical components, equipment and machinery are also very prone to damage from water.

When locating components and equipment, therefore, the humidity of the air and the possible formation of water particles must always be taken into consideration, particularly as humidity possesses (almost without exception) a certain amount of electrical conductivity which increases the possibility of corrosion of metals. Similarly, the ingress of water followed by freezing within machinery can result in malfunction.

1.8.1 Salt water

Salt has an electrochemical effect on metallic materials (i.e. corrosion) which can damage and degrade the performance of equipment and/or parts that have been manufactured from metallic materials. Non-metallic materials can also be damaged by salt through a complex chemical reaction which is dependent on the supply of oxygenated salt solution to the surface of the material, its temperature, and the temperature and humidity of the environment. This is particularly a problem in areas close to the sea or mountain ranges.

1.8.2 Ice and snow

Water in the form of ice can cause problems in the cooling of equipment (or by freezing and then thawing) which will result in cracks occurring, cases breaking, etc. Powdered snow can easily be blown through ventilation ducts and then melt down in equipment compartments and cubicles, which can cause damp problems to critical systems if not prevented in the original construction.

1.8.3 Weathering

'Weathering' is the collective term for the processes by which rock at or near the Earth's surface is disintegrated and decomposed by the action of atmospheric agents, water and living things. Its effect on equipment and materials must always be considered. There are several types of weathering, as shown below.

Exfoliation

Rocks are gradually warmed by the sun during the day, but at night the surface will cool far more rapidly than the underlying rocks. The outer skin of the rock will then become tight and crack, and layers of the rock will peel off, making the rock face rounded or dome shaped. This exfoliation can affect equipment that is sunk into the rock face or mounted on the surface of the rock.

Freeze–thaw

When water freezes it turns to ice, expanding by about one-twelfth of its volume. If this water is in a joint or a crack in equipment casing then that space will become enlarged and the casing on either side will be forced apart. When the ice eventually thaws more water will penetrate the crack and the cycle will repeat itself, with the crack constantly enlarging.

Chemical weathering

Water can pick up quantities of sulphur dioxide from the atmosphere, forming a weak solution of acid that can attack certain equipment housings, which may become worn away.

Erosion

As the wind blows over dry ground it collects grit and throws it vigorously against nearby surfaces. This grit acts in a similar way to sandpaper, gradually wearing away the surface of equipment or objects with which it comes into contact.

Mass movement

Once solids such as sedimentary rocks have been broken up, there is often a downwards movement of the particles that have broken off. This 'soil creep' can gather momentum and can, in certain circumstances, submerge equipment.

1.8.4 Typical requirements: weather and precipitation

The most common environmental requirements concerning weather and precipitation are shown in Table 1.2.

1.9 What is the Water Regulations Advisory Scheme?

The Water Regulations Advisory Scheme (WRAS) is a joint venture between WRc plc and NSF International (WRc-NSF) and is funded by all of the UK's water supply companies. Its main aim is to promote knowledge of the Water

Table 1.2 **Typical requirements: water and precipitation**

Weather protection	Equipment that is operated adjacent to the seashore or on mountain ranges (and therefore subject to water and precipitation) must be able to function equally well as the same equipment housed in arid deserts
Operation	All equipment should be capable of operating during rain, snow and hail, and be unaffected by ice, salt and water
Rain	All equipment should be capable of operating in rain and be capable of preventing the penetration of rainfall at a minimum rate of 13 cm/h and an accompanying wind rate of 25 m/s
Snow and hail	Consideration needs to be given to the effect of all forms of snow and/or hail. The maximum diameter of the hailstones is conventionally taken as 15 mm, but larger diameters can occur on occasions
Salt water	Equipment should be capable of operating in (or be protected from) heavy salt spray, as would be experienced in seacoast areas and in the vicinity of salted roadways

Regulations (i.e. the Water Supply (Water Fittings) Regulations 1999, the Water Bye-laws 2004 (Scotland) and the Northern Ireland Water Regulations) throughout the UK and to encourage their consistent interpretation and enforcement, for the prevention of waste, undue consumption, misuse, erroneous measurement or contamination of water.

The WRAS:

- Provides a postal, telephone and e-mail advisory service on the requirements of the Water Regulations (i.e. the Water Supply (Water Fittings) Regulations 1999, the Water Bye-laws 2004 (Scotland) and the Northern Ireland Water Regulations). The service is provided to water suppliers and any other persons or bodies who require guidance on the principles of the Regulations and Bye-laws.
- Arranges for testing of water fittings and materials to ascertain whether they comply with the current UK legislation.
- Publishes the *Water Regulations Guide*, which (according to the WRAS) 'is of inestimable value to those designing fittings, making and installing fittings and controlling the installation of water fittings'.
- Publishes the *Water Fittings and Materials Directory*, which lists approved fittings and materials for use on the UK water supply system.
- Publishes information and guidance notes on various topics relating to the Regulations and Bye-laws (e.g. Precautions against freezing, Type BA (RPZ valve) backflow preventers, Reclaimed water systems).
- Publishes guidance booklets aimed at specific types of premises on how to ensure that water installations comply with the Regulations and Bye-laws (e.g. Dental practices, Agricultural and Photographic premises).
- Represents the UK water supply industry on many national and international standards committees relating to the supply of water.

In addition, the WRAS administers a scheme to assess, register and publicize approved contractors who work in accordance and compliance with the Water Regulations.

The WRAS has over 75 years experience of the water supply industry, helping supply companies, regulators and consumers to obtain wholesome water at the tap. With purpose-built testing and calibration laboratories in Reading, Berkshire and Oakdale, Gwent, it operates the water industry's scheme for approving water fittings and materials for compliance with the regulations, as well as providing information to support the design, build and acceptance of materials, fittings and products used in water supply and sewerage.

The WRc-NSF is recognized as one of the world's largest test laboratories for water fittings, having expertise in all aspects of mechanical and hygienic requirements for products used for the treatment, supply and use of water, and is capable of carrying out tests on a large number of different types of fixtures and fittings to assist manufacturers in product development, including:

- backflow prevention devices
- gate, solenoid and spherical valves
- pipes and pipe systems
- pressure and temperature relief valves
- reduced pressure zone (RPZ) valves
- tanks and cisterns
- taps, water outlets and mixers
- toilet suites
- thermostatic mixing valves
- unvented hot water apparatus
- water meters.

The laboratory can also carry out a wide range of tests in both hot and cold water to assist manufacturers in product development, including:

- extraction of metals testing
- extraction of substances of concern to human health (cytotoxicity) testing
- growth of biofilm testing
- identification of unknown or unexpected trace substances leaking from materials
- odour and flavour assessment
- total organic carbon leaching.

The WRAS also carries out routine testing for:

- WRAS Approval to BS 6920
- System Assurance Institute (SAI) Global Approval to AS/NZS 4020
- National Sanitation Foundation (NSF) Approval to NSF/ANSI 61
- UK Secretary of State (Drinking Water Inspectorate) Approval
- Testing to the BS 12873 series of standards.

The results from WRc-NSF test laboratories are accepted by a wide range of certification and approval bodies around the world.

For more information about the WRAS, see their website at www.wras. co.uk or contact them at: Water Regulations Advisory Scheme, Fern Close, Pen-y-Fan Industrial Estate, Oakdale, Gwent NP11 3EH. Tel: +44 (0)1495 248454. Fax: +44 (0)1495 249234. E-mail: info@wras.co.uk

2

What are the Water Regulations?

In England and Wales, the Water Supply (Water Fittings) Regulations 1999 (Statutory Instrument 1999 No. 1148 which came into force on 1st July 1999) replaced the Bye-laws previously issued by water companies.

In Scotland, the Scottish Water Bye-laws 2004 came into force on 30th August 2004 and replaced the previous Water Bye-laws 2000, Scotland. These 2004 Bye-laws mirror the requirements of the English and Welsh 1999 Regulations (other than Bye-laws 8 and 10, which refer to legislation that is not applicable to Scotland) and, for consistency, use the same paragraph numbering system.

There are, however, some slight differences in the text to reflect national differences and organizational titles. For example, in England and Wales the authority responsible for overseeing conformance to the Regulations/Bye-laws is called the 'water undertaker', while in Scotland the authority is called the 'undertaker' (to avoid confusion, hereafter this organization shall be referred to in this book as the 'water undertaker'). Where these differences occur, they have been highlighted when they first appear in the text.

Note: Northern Ireland is slightly different in that Northern Ireland Water is a government-owned company (GoCo), which is a statutory trading body owned by central government but operating under company legislation, with substantial independence from government. They have offices across Northern Ireland to deliver the services that Northern Ireland Water provides and work in conformance with Bye-laws that are very similar to the English, Scottish and Welsh Regulations and Bye-laws. They are currently in the process of preparing new regulations to ensure that the standards and requirements for water fittings used in Northern Ireland are broadly similar to those contained in the Water Supply (Water Fittings) Regulations 1999.

2.1 What is the aim of the Regulations/ Bye-laws?

The Water Regulations set requirements for the design, installation and maintenance of plumbing systems and water fittings and are enforced by water companies in their respective areas of supply. Water companies, therefore, have a legal obligation to ensure that they have been implemented, including the inspection of new and existing installations.

These Regulations/Bye-laws apply to any water fitting installed or used (or to be installed and/or used) in premises to which water is (or is to be) supplied by a water undertaker. They are aimed at preventing waste, undue consumption, misuse and/or contamination of water supplied by a water undertaker.

These Regulations/Bye-laws do *not* apply to:

- certain water fittings in connection with water supplied for non-domestic or food production purposes, *neither* do they require any person to remove, replace, alter, disconnect or cease to use any water fitting which was lawfully installed or used, or capable of being used, before the current Regulations and Bye-laws came into force;
- a water fitting installed or used in connection with water supplied for purposes other than domestic or food production purposes, provided that:
  - the water is metered;
  - the supply of the water is for a period not exceeding one month; and
  - no water can return through the meter to any pipe that is the responsibility of the water undertaker;
- water fittings which are not connected to water supplied by a water undertaker.

2.2 How do the Regulations affect you?

Water systems and fittings in premises that are, or will be, connected to the public water supply must comply with the Regulations. If you are planning to carry out certain plumbing work, you must obtain the prior consent of your water company by giving advance notice of the work. This includes the installation of water fittings in connection with the:

- construction of a large pond or swimming pool with automatic replenishment;
- erection of any new building or structure;
- extension or alteration of the water system in any premises except a domestic dwelling
- material change in use of any premises;
- installation of any fitting listed in section 5 of the Regulations.

2.3 What is their content?

The Regulations/Bye-laws are split into five parts, as shown in Table 2.1.

Table 2.1 Content of the Water Supply (Water Fittings) Regulations/Bye-laws

Part	Title	No.	Regulation/Bye-law	Content	Remarks
Part I	Preliminary	1	Citation, commencement and interpretation	Glossary of terms	
		2	Application of Regulations/Bye-laws	Aim of the Regulations/Bye-laws	
Part II	Requirements	3	Restriction on installation, etc., of water fittings	Regulation/Bye-laws 3 and 4 impose general requirements in relation to water fittings	Water fittings must not be installed, connected, arranged or used in such a manner that they are likely to cause waste, misuse, undue consumption or contamination, or erroneous measurement, of the water supplied. They must be of an appropriate quality or standard, and be suitable for the circumstances in which they are used; and they must be installed, connected or disconnected in a workmanlike manner
		4	Requirements for water fittings, etc.		
		5	Notification	Regulation/Bye-law 5 requires a person who proposes to install certain water fittings to notify the undertaker, and not to commence installation without the undertaker's consent	The water undertaker may withhold consent or grant it on certain conditions. This requirement does not apply to some fittings which are installed by a contractor who is approved by the water undertaker or certified by an organization specified by the Regulator. Where an approved contractor installs, alters, connects or disconnects a water fitting, he must provide a certificate stating whether it complies with the Regulations/Bye-laws
		6	Contractor's certificate	Regulation/Bye-law 6 describes the contractor's liability for providing a certificate following the installation of a water fitting	
Part III	Enforcement, etc.	7	Penalty for contravening Regulations/Bye-laws	Regulation/Bye-law 7 provides a fine not exceeding level 3 (currently £1000) in England and Wales) or level 5 (currently £5000) in Scotland for contravening the Regulations/Bye-laws	It is a defence to show that the work on a water fitting was done by (or under the direction of) an approved contractor, and that the contractor certified that it complied with the Regulations/Bye-laws. In England and Wales, this defence is extended to the offences of contaminating, wasting and misusing water under section 73 of the Water Industry Act 1991
		8	Modification to section 73 of the Water Industry Act 1991	Amendment to the Water Industry Act 1991 concerning contamination, wasting and misusing water, etc.	This Regulation only applies to England and Wales

9	Inspections, measurements and tests	Regulation/Bye-law 9 enables water undertakers and local authorities who are required to enter premises to carry out inspections, measurements and tests for the purposes of the Regulations/Bye-laws
10	Enforcement	Regulation/Bye-law 10 requires the water undertaker to enforce the requirements of the Regulations/Bye-laws	This Regulation only applies to England and Wales and is enforceable by the Regulator or the Director General of Water Services
11	Relaxation of Requirements	Regulation/Bye-law 11 enables the water undertaker to apply for a moderation of certain requirements	In England and Wales, where the water undertaker considers that any requirement of these Requirements/Bye-laws would be inappropriate in relation to a particular case or class of cases, the water undertaker may apply to the Regulator. In Scotland the water undertaker would apply to the Scottish Ministers
12	Approval by the water undertaker	Where the water undertaker approves a method of installation, the water undertaker shall give notice of approval to the Regulator (in England and Wales) and to the Scottish Ministers (in Scotland)	In England and Wales, Regulations 12 requires the Regulator to consult water undertakers and organizations representing water users before giving an approval for the purpose of the Regulation, and to publicize approvals
13	Disputes	Regulation/Bye-law 13 provides for disputes arising under the Regulations/Bye-laws between a water undertaker and a person who has installed (or proposes to install) a water fitting to be referred to arbitration	
14	Revocation of Bye-laws	Regulation/Bye-law 14 revokes the existing water Bye-laws made by water undertakers	This refers to: • English and Welsh Bye-laws made under section 17 of the Water Act 1945 • Scottish Bye-laws made under section 70 of the Water (Scotland) Act 1980

2.4 What are the essential requirements?

Every water fitting *shall*:

- be of an appropriate quality and standard;
- be suitable for the circumstances in which it is used;
- comply with the requirements of Schedule 2 to these Regulations/Bye-laws;
- be installed, connected, altered, repaired or disconnected in a workman-like manner;
- bear an appropriate CE marking;
- conform to an appropriate harmonized standard or specification approved by the Regulator.

In Scotland, a water fitting must also conform to the terms of a specification approved under the Water Supply (Water Fittings) Regulations 1999.

2.5 Installation of water fittings

- No person shall install a water fitting to convey or receive water supplied by a water undertaker, or alter, disconnect or use such a water fitting.
- No water fitting shall be installed, connected, arranged or used in such a manner that it causes or is likely to cause:
  - waste, misuse, undue consumption or contamination of water supplied by a water undertaker; or
  - the erroneous measurement of water supplied by a water undertaker.
- No water fitting shall be installed, connected, arranged or used if it is damaged, worn or otherwise faulty and is likely to cause:
  - waste, misuse, undue consumption or contamination of water supplied by a water undertaker; or
  - the incorrect measurement of water supplied by a water undertaker.

2.6 Notification by persons proposing to install a water fitting

Any person proposing to install a water fitting in connection with any of the operations listed in Table 2.2:

- *shall* give notice to the water undertaker that he proposes to begin work;
- *shall not* begin that work without the consent of that water undertaker; and
- *shall* comply with any conditions made by the water undertaker.

Table 2.2 **Installations of water fittings requiring notification**

Serial	Type of installation
1	The erection of a building or other structure that is not a pond or swimming pool (but see Serial 5)
2	The extension or alteration of a water system on any premises other than a house
3	The material change of use of any premises
4	The installation of: (a) a bath with a capacity greater than 230 litres; (b) a bidet with an ascending spray or flexible hose; (c) a single shower unit (as opposed to a drench shower installed for reasons of safety or health) that is connected directly or indirectly to a supply pipe; (d) a pump or booster drawing more than 12 litres per minute, connected directly or indirectly to a supply pipe; (e) a unit which incorporates reverse osmosis; (f) a water treatment unit which discharges wastewater or needs water for regeneration and/or cleaning; (g) a reduced pressure zone valve assembly (or other mechanical device) for protection against a fluid which is in fluid category 4 or 5 (see subsection 2.14 and Appendix 1 to section 2); (h) a garden watering system (unless hand operated); or (i) any water system that is laid outside a building and is either less than 750 mm or more than 1350 mm below ground level
5	The construction of a pond or swimming pool with a capacity greater than 10,000 litres which is designed to be automatically replenished and filled with water supplied by a water undertaker

The notice *shall* include or be accompanied by:

- the name and address of the person giving the notice;
- a description of the proposed work or material change of use;
- the location of the premises to which the proposal relates together with the use (or intended use) of those premises;
- a plan of those parts of the premises to which the proposal relates plus a diagram showing the pipework and fitting to be installed (unless the installation is of the type listed in 4a, 4c, 4h or 5 in Table 2.2).

Note: The water undertaker may withhold their consent or grant it subject to conditions.

2.7 Contractor's certificate

Where a water fitting is installed, altered, connected or disconnected by an approved contractor, the contractor shall upon completion of the work provide a signed certificate stating whether the water fitting complies with the requirements of these Regulations/Bye-laws, to the person who commissioned the work.

And to the water undertaker if the installation is of a type 5 shown in Table 2.2.

2.8 Enforcement

The duty of a water undertaker is enforceable under section 18 of the Water Industry Act 1991.

A water undertaker shall enforce the requirements of these Regulations/ Bye-laws in relation to the area for which it holds an appointment under Part II of the Water Industry Act 1991.

2.9 Approval of the water undertaker

Where the water undertaker approves a method of installation, he shall give notice of the approval to the Regulator/Scottish Minister (as applicable) and shall publish it in such manner as the undertaker considers appropriate.

 **Note:** In England and Wales, before approving a specification and/or a requirement for a water fitting, the Regulator shall consult

- every water undertaker;
- such trade associations as he considers appropriate; and
- such organizations appearing to him to be concerned with the interests of water users as he considers appropriate.

Where the Regulator approves a specification he shall give notice of the approval to all persons who were consulted and shall publish it in such manner as he considers appropriate.

2.10 Penalty for contravening Regulations/ Bye-laws

A person who:

- contravenes the requirements for the installation of water fittings or fails to provide a contractor's certificate, or
- commences an operation listed in Table 2.2 without giving the necessary notice or consent;

is guilty of an offence and liable on summary conviction to a fine not exceeding level 3 (currently £1000) in England and Wales or level 5 (currently £5000) in Scotland for contravening the Regulations/Bye-laws.

2.11 Inspections, measurements and tests

A water undertaker may designate (in writing) a person(s) to carry out inspections, measurements and tests:

- on premises entered by that person; or
- on water fittings;
- or other articles found on any such premises; and

to take away such samples of water or of any land, and such water fittings and other articles, as that person may consider necessary for further examination and or evidence.

2.12 Relaxation of requirements

Where a water undertaker considers that any requirement of these Regulations/ Bye-laws would be inappropriate in relation to a particular case, the water undertaker may apply to the Regulator (in England and Wales) or to the Scottish Ministers (in Scotland) to authorize a relaxation of that requirement.

2.13 Disputes

Any dispute between a water undertaker and a person who has installed or proposes to install a water fitting as to whether the water undertaker has unreasonably:

- withheld consent; or
- attached unreasonable conditions; or
- refused to apply to the Regulator for a relaxation of the requirements;

shall be referred to arbitration by a single arbitrator to be appointed by agreement between the parties or, in default of agreement, by the Regulator or Scottish Ministers as applicable.

2.14 What are the fluid categories?

The fluid categories are described in Table 2.3.

Table 2.3 **Fluid categories**

Fluid category	Description	Example	Remarks
1	Wholesome water supplied by a water undertaker and complying with the requirements of regulations made under water section 67 of the Water Industry Act 1991	• Water supplied directly from a water undertaker's main	Wherever practicable, water for drinking purposes should be obtained directly from a supply pipe, that is, without any intervening storage before use
2	Water in fluid category 1 whose aesthetic quality is impaired owing to: • a change in its temperature; or • the presence of substances or organisms causing a change in its taste, odour or appearance, including water in a hot water distribution system	• Water heated in a hot water secondary system • Mixtures of fluid categories 1 and 2 water discharged from combination taps or showers • Water that has been softened by a domestic common salt regeneration process • Mixing of hot and cold water supplies • Domestic softening plant (common salt regeneration) • Drink vending machines in which no ingredients or carbon dioxide are injected into the supply or distributing inlet pipe • Fire sprinkler systems (without antifreeze) • Ice-making machines • Water-cooled air-conditioning units (without additives)	These changes in water quality are aesthetic changes only and the water is considered to present no human health hazard

| 3 | Fluid which represents a slight health hazard because of the concentration of substances of low toxicity, including any fluid which contains:
• ethylene glycol;
• copper sulphate solution, or similar chemical additives; or
• sodium hypochlorite (chloros and common disinfectants) | • Water in primary circuits and heating systems (with or without additives) in a house
• Domestic washbasins, baths and showers
• Domestic clothes and dishwashing machines
• Home dialysing machines
• Drink vending machines in which ingredients or carbon dioxide are injected
• Commercial softening plant (common salt regeneration only)
• Domestic hand-held hoses with flow controlled spray or shut-off control
• Hand-held fertilizer sprays for use in domestic gardens
• Domestic or commercial irrigation systems, without insecticide or fertilizer additives and with fixed sprinkler heads not less than 150 mm above ground level

and

• In premises other than single-occupancy dwellings;

Note: Where domestic fittings, washbasins, baths or showers are installed in premises other than single-occupancy dwellings, that is, commercial, industrial or other premises, these appliances may still be regarded as fluid category 3, unless there is a potential higher risk. Typical premises in which some, or all, of these appliances may be regarded as justifying a higher fluid risk category include hospitals and other medical establishments | Category 3 represents a slight health hazard and is not suitable for drinking or other domestic purposes |

(Continued)

Table 2.3 Continued

Fluid category	Description	Example	Remarks
4	Fluid which represents a significant health hazard due to the concentration of toxic substances, including any fluid which contains: • chemical, carcinogenic substances or pesticides (including insecticides and herbicides); or • environmental organisms of potential health significance	• Water in primary circuits and heating systems other than in a house, irrespective of whether additives have been used or not • Water treatment or softeners using than salt • Clothes and dishwashing machines for other than domestic use • Mini-irrigation systems in house gardens without fertilizer or insecticide application such as pop-up sprinklers, permeable hoses, or fixed or rotating sprinkler heads fixed less than 150 mm above ground level General • Primary circuits and central heating systems other than in a house • Fire sprinkler systems using antifreeze solutions • House gardens • Mini-irrigation systems without fertilizer or insecticide application; such as pop-up sprinklers or permeable hoses • Food processing • Food preparation • Dairies • Catering • Commercial dishwashing machines • Bottle-washing apparatus • Refrigerating equipment • Industrial and commercial installations	Category 4 represents a significant health hazard and is not suitable for drinking or other domestic purposes

5

Fluid representing a serious health hazard because of the concentration of pathogenic organisms, radioactive or very toxic substances, including any fluid which contains:

- faecal material or other human waste; or
- butchery or other animal waste; or
- pathogens from any other source

- Dyeing equipment
- Industrial disinfection equipment
- Printing and photographic equipment
- Car washing and degreasing plants
- Commercial clothes washing plants
- Brewery and distillation plant
- Water treatment plant or softeners using other than salt
- Pressurized fire-fighting systems

- Sinks, urinals, WC pans and bidets in any location
- Permeable pipes or hoses other than in domestic gardens, laid below or at ground level, with or without chemical additives
- Grey water recycling systems
- Clothes and dishwashing machines in high-risk premises

General

- Industrial cisterns
- Non-domestic hose union taps
- Sinks, urinals, WC pans and bidets
- Permeable pipes in other than domestic gardens, laid below or at ground level, with or without chemical additives
- Grey water recycling systems

Medical

- Any medical or dental equipment with submerged inlets
- Laboratories

Fluid category 5 represents a serious health hazard and is the most polluting category listed

(Continued)

Table 2.3 Continued

Fluid category	Description	Example	Remarks
		• Bedpan washers	
		• Mortuary and embalming equipment	
		• Hospital dialysing machines	
		• Commercial clothes washing plant in healthcare premises	
		• Non-domestic sinks, baths, washbasins and other appliances	
		• Food processing	
		• Butchery and meat trades	
		• Slaughterhouse equipment	
		• Vegetable washing	
		• Dishwashing machines in health care premises	
		• Industrial and commercial installations	
		• Industrial and chemical plant, etc.	
		• Mobile plant, tankers and gully emptiers	
		• Laboratories	
		• Sewage treatment and sewer cleansing	
		• Drain cleaning plant	
		• Water storage for agricultural purposes	
		• Water storage for fire-fighting purposes	
		• Commercial agricultural	
		• Commercial irrigation outlets below or at ground level and/or permeable pipes, with or without chemical additives	
		• Insecticide or fertilizer applications	
		• Commercial hydroponic systems	

2.15 What are the mandatory requirements for water fittings?

No material or substance, either alone or in combination with any other material or substance or with the contents of any water fitting of which it forms a part, which causes or is likely to cause contamination of water shall be used in the construction, installation, renewal, repair or replacement of any water fitting which conveys or receives, or may convey or receive, water supplied for domestic or food production purposes.

This requirement does not, however, apply to a water fitting downstream of a terminal fitting supplying wholesome water where:

- the use to which the water downstream is put does not require wholesome water; and
- a suitable arrangement or device to prevent backflow is installed.

2.15.1 General

Every water fitting *shall*:

• be immune to or be protected from corrosion by galvanic action or by any other process which is likely to result in contamination or waste of water;	SI 1148-3(a)
• be constructed of materials of such strength and thickness as to resist damage from any external load, vibration, stress or settlement, pressure surges, or temperature fluctuation to which it is likely to be subjected;	SI 1148-3(b)
• be watertight;	SI 1148-4(a)
• be so constructed and installed as to: ○ prevent ingress by contaminants; and ○ inhibit damage by freezing or any other cause;	SI 1148-4(b)
• be so installed as to minimize the risk of permeation by, or deterioration from contact with, any substance which may cause contamination;	SI 1148-4(c)
• be adequately supported;	SI 1148-4(d)
• be capable of withstanding an internal water pressure not less than 1½ times the maximum pressure to which that fitting is designed to be subjected in operation.	SI 1148-5

No water fitting *shall*:

- be installed, connected or used which is likely to have a detrimental effect on the quality or pressure of water in a water main or other pipe of a water undertaker;   SI 1148-6
- be embedded in any wall or solid floor.   SI 1148-7(1)

No water fitting which is designed to be operated or maintained, whether manually or electronically, or which consists of a joint, shall be a concealed water fitting.   SI 1148-7(2)

Any concealed water fitting or mechanical backflow prevention device, not being a terminal fitting, shall be made of gunmetal, or another material resistant to dezincification.   SI 1148-7(3)

Any water fitting laid below ground level shall have a depth of cover sufficient to prevent water freezing in the fitting.   SI 1148-7(4)

Note: 'Concealed water fitting' means a water fitting which:

- is installed below ground;
- passes through or under any wall, footing or foundation;
- is enclosed in any chase or duct; or
- is in any other position which is inaccessible or renders access difficult.

2.15.2 Design and installation

No water fitting shall be installed in such a position, or pass through such surroundings, that it is likely to cause contamination or damage to the material of the fitting or the contamination of water supplied by the water undertaker.   SI 1148-8

Every supply pipe or distributing pipe providing water to separate premises *shall* be fitted with a stopvalve conveniently located to enable the supply to those premises to be shut off without shutting off the supply to any other premises.   SI 1148-10(1)

Where a supply pipe or distributing pipe provides water in common to two or more premises, it *shall* be fitted with a stopvalve to which each occupier of those premises has access.   SI 1148-10(2)

Any pipe supplying cold water for domestic purposes to SI 1148-9
any tap *shall* be so installed that, so far as is reasonably
practicable, the water is not warmed above 25°C.

Water supply systems *shall* be capable of being drained SI 1148-11
down and be fitted with an adequate number of servicing
valves and drain taps so as to minimize the discharge of
water when water fittings are maintained or replaced.
A sufficient number of stopvalves *shall* be installed for
isolating parts of the pipework.

The water system *shall* be capable of withstanding SI 1148-12(1)
an internal water pressure not less than 1½ times the
maximum pressure to which the installation or relevant
part is designed to be subjected in operation (i.e. 'the
test pressure').

This requirement shall be deemed to be satisfied: SI 1148-12(2)

- in the case of a water system that does not include
  a pipe made of plastics, where:
  - the whole system is subjected to the test pressure
    by pumping, after which the test continues for
    one hour without further pumping;
  - the pressure in the system is maintained for one
    hour; and
  - there is no visible leakage throughout the test;
- in any other case, where either of the tests shown
  above is satisfied.

2.15.3 Prevention of cross-connection to unwholesome water

Any water fitting conveying: SI 1148-14(1)(a)

- rainwater, recycled water, or any fluid other
  than water supplied by a water undertaker; or
- any fluid that is not wholesome water; 14(1)(b)

shall be clearly identified so as to be easily
distinguished from any supply pipe or distributing pipe.

No supply pipe, distributing pipe or pump delivery pipe drawing water from
a supply pipe or distributing pipe shall convey, or be connected so that it can
convey, any fluid falling within sub-paragraph 14(1) (see above) unless a
device for preventing backflow is installed with sub-paragraph 15.

2.15.4 Backflow prevention

Every water system shall contain an adequate device or devices for preventing backflow of fluid from any appliance, fitting or process from occurring unless it is:	SI 1148-15(1) 15(2)(a)
• a water heater where the expanded water is permitted to flow back into a supply pipe; or	
• a vented water storage vessel supplied from a storage cistern;	15(2)(b)
where the temperature of the water in the supply pipe or the cistern does not exceed 25°C.	
The device used to prevent backflow shall be appropriate to the highest applicable fluid category to which the fitting is subject downstream before the next such device.	SI 1148-15(3)
Backflow prevention *shall* be provided on any supply pipe or distributing pipe:	SI 1148 15(4)(a)
• where it is necessary to prevent backflow between separately occupied premises, or	
• where the water undertaker has given notice for the purposes of this Schedule that such prevention is needed for the whole or part of any premises.	15(4)(b)

 Note: A backflow prevention device is adequate for the purposes of paragraph 15(1) if it is in accordance with a specification approved by the Regulator/Scottish Ministers (as applicable) for the purposes of this Schedule.

2.15.5 Cold water storage cisterns

Every inlet to a storage cistern, combined feed and expansion cistern, WC flushing cistern or urinal flushing cistern *shall* be fitted with a servicing value on the inlet pipe adjacent to the cistern.	SI 1148-16(2)
Every storage cistern, except one supplying water to the primary circuit of a heating system, *shall* be fitted with a servicing valve on the outlet pipe.	SI 1148-16(3)

Every storage cistern *shall* be fitted with: SI 1148
                                                                  16(4)(a)

- an overflow pipe, with a suitable means of
  warning of an impending overflow, which
  excludes insects;
- a cover positioned so as to exclude light and   16(4)(b)
  insects;
- thermal insulation to minimize freezing or   16(4)(b)
  undue warming.

Every storage cistern *shall* be so installed as to   SI 1148-16(5)
minimize the risk of contamination of stored water.
The cistern shall be of an appropriate size, and the
pipe connections to the cistern shall be so positioned,
as to allow free circulation and to prevent areas of
stagnant water from developing.

Every pipe supplying water connected to a storage   SI 1148-16(1)
cistern *shall* be fitted with an effective adjustable valve
capable of shutting off the inflow of water at a suitable
level below the overflowing level of the cistern.

2.15.6 Hot water heating services

Every unvented water heater, not being an instantaneous SI 1148-
water heater with a capacity not greater than 15 litres, 17(1)(a)
and every secondary coil contained in a primary system
shall:

- be fitted with a vent pipe, a temperature control
  device, a temperature relief device and a combined
  temperature pressure and relief valve; or
- be capable of accommodating expansion within   17(1)(b)
  the secondary hot water system.

An expansion valve *shall* be fitted with provision to   SI 1148-17(2)
ensure that water is discharged in a correct manner in
the event of a malfunction of the expansion vessel or
system.

Every expansion valve *shall*:   SI 1148
                                                          22(2)(a)
- be fitted on the supply pipe close to the hot water
  vessel and without any intervening valves; and

- only discharge water when subjected to a water pressure of not less than 0.5 bar (50 kPa) above the pressure to which the hot water vessel is, or is likely to be, subjected in normal operation.   22(2)(b)

Every expansion valve, temperature relief valve or combined temperature and pressure relief valve connected to any fitting or appliance *shall* close automatically after a discharge of water.   SI 1148-22(1)

Every expansion cistern or expansion vessel, and every cold water combined feed and expansion cistern connected to a primary circuit, *shall* be such as to accommodate any expansion water from that circuit during normal operation.   SI 1148-21

Discharges from temperature relief valves, combined temperature and pressure relief valves and expansion valves *shall* be made in a safe and conspicuous manner.   SI 1148-19

A temperature relief valve or combined temperature and pressure relief valve *shall* be provided on every unvented hot water storage vessel with a capacity greater than 15 litres.   SI 1148-23(1)

The valve *shall*:   SI 1148
                                                            23(2)(a)

- be located directly on the vessel in an appropriate location, and have a sufficient discharge capacity, to ensure that the temperature of the stored water does not exceed 100°C; and
- only discharge water at below its operating temperature when subjected to a pressure of not less than 0.5 bar (50 kPa) in excess of the greater of the following: 23(2)(b)
  - the maximum working pressure in the vessel in which it is fitted, or
  - the operating pressure of the expansion valve.

Appropriate vent pipes, temperature control devices and combined temperature and pressure relief valves *shall* be provided to prevent the temperature of the water within a secondary hot water system from exceeding 100°C.   SI 1148-18

No vent pipe from a primary circuit *shall* terminate over a storage cistern containing wholesome water for domestic supply or for supplying water to a secondary system.   SI 1148-20(1)

No vent pipe from a secondary circuit *shall* terminate over any combined feed and expansion cistern connection to a primary circuit.   SI 1148-20(2)

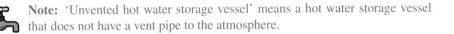

Note: 'Unvented hot water storage vessel' means a hot water storage vessel that does not have a vent pipe to the atmosphere.

> No supply pipe or secondary circuit shall be permanently connected to a closed circuit for filling a heating system unless it incorporates a backflow prevention device in accordance with a specification approved by the Regulator/Scottish Ministers (as applicable) for the purposes of this Schedule.
>
> SI 1148-24

2.15.7 Water closets and flushing devices

> Every water closet pan *shall* be supplied with water from a flushing cistern, pressure flushing cistern or pressure flushing valve, and shall be so made and installed that after normal use its contents can be cleared effectively by a single flush of water, or, where the installation is designed to receive flushes of different volumes, by the largest of those flushes.
>
> SI 1148 25(1)(a)
>
> Every water closet, and every flushing device designed for use with a water closet, shall comply with a specification approved by the Regulator/ Scottish Ministers for the purposes of this Schedule.
>
> SI 1148-25(2)
>
> No pressure flushing valve *shall* be installed –
>
> SI 1148 25(1)(b)
>
> * in a house, or
> * in any building not being a house
>
> where a minimum flow rate of 1.2 litres per second cannot be achieved at the appliance.
>
> Where a pressure flushing valve is connected to a supply pipe or distributing pipe, the flushing arrangement *shall* incorporate a backflow prevention device consisting of a permanently vented pipe interrupter located not less than 300 mm above the spillover level of the WC pan or urinal.
>
> SI 1148 25(1)(c)
>
> No flushing device installed for use with a WC pan shall give a single flush exceeding 6 litres.
>
> SI 1148 25(1)(d)

Note: Notwithstanding sub-paragraph 25(1)(d) a flushing cistern installed before 1st July 1999 may be replaced by a cistern which delivers a similar volume and which may be either single flush or dual flush; but a single flush cistern may not be so replaced by a dual flush cistern.

No flushing device designed to give flushes of different volumes shall have a lesser flush exceeding two-thirds of the largest flush volume.	SI 1148 25(1)(f) SWB 25(1)(e)

A flushing device designed to give flushes of different volumes:

SI 1148 25(1)(h)

- shall have a readily discernible method of actuating the flush at different volumes; and
- shall have instructions, clearly and permanently marked on the cistern or displayed nearby, for operating it to obtain the different volumes of flush.

SWB 25(1)(g)

Every flushing cistern, other than a pressure flushing cistern, *shall* be clearly marked internally with an indelible line to show the intended volume of flush, together with an indication of that volume.

SI 1148 25(1)(g)

SWB 25(1)(f)

Every flushing cistern, not being a pressure flushing cistern or a urinal cistern, *shall* be fitted with a warning pipe or with a no less effective device.

SI 1148 25(1)(i)
SWB 25(1)(h)

Every urinal that is cleared by water after use *shall* be supplied with water from a flushing device which:

- in the case of a flushing cistern, is filled at a rate suitable for the installation;
- in all cases, is designed or adapted to supply no more water than is necessary for effective flow over the internal surface of the urinal and for replacement of the fluid in the trap.

SI 1148 25(1)(j)

SWB 25(1)(i)

Except in the case of a urinal which is flushed manually, or which is flushed automatically by electronic means after use, every pipe which supplies water to a flushing cistern or trough used for flushing a urinal shall be fitted with an isolating valve controlled by a time switch and a lockable isolating valve, or with some other equally effective automatic device for regulating the periods during which the cistern may fill.

SI 1148 25(1)(k)

SWB 25(1)(j)

Note: The requirement in sub-paragraph 25(1)(i) *shall* be deemed to be satisfied:

- in the case of an automatically operated flushing cistern servicing urinals which is filled with water at a rate not exceeding:
  - 10 litres per hour for a cistern serving a single urinal;
  - 7.5 litres per hour per urinal bowl or stall, or, as the case may be, for each 700 mm width of urinal slab, for a cistern serving two or more urinals;
- in the case of a manually or automatically operated pressure flushing valve used for flushing urinals which delivers not more than 1.5 litres per bowl or position each time the device is operated.

The requirements of sub-paragraphs 25(1) and 25(2) do not apply where faeces or urine are disposed of through an appliance that does not solely use fluid to remove the contents.

2.15.8 Baths, sinks, showers and taps

All premises supplied with water for domestic purposes *shall* have at least one tap conveniently situated for the drawing of drinking water.	SI 1148-26
A drinking water tap *shall* be supplied with water from:	SI 1148 27(a)
- a supply pipe; or	
- a pump delivery pipe drawing water from a supply pipe; or	27(b)
- a distributing pipe drawing water exclusively from a storage cistern supplying wholesome water.	27(c)
Every bath, wash basin, sink or similar appliance *shall* be provided with a watertight and readily accessible plug or other device capable of closing the waste outlet.	SI 1148-28(1)

Note: This requirement does not apply to:

- an appliance where the only taps provided are spray taps;
- a washing trough or wash basin whose waste outlet is incapable of accepting a plug and to which water is delivered at a rate not exceeding 0.06 litres per second exclusively from a fitting designed or adapted for that purpose;
- a wash basin or washing trough fitted with self-closing taps;
- a shower bath or shower tray;
- a drinking water fountain or similar facility; or
- an appliance which is used in medical, dental or veterinary premises and is designed or adapted for use with an unplugged outlet.

2.15.9 Washing machines, dishwashers and other appliances

Clothes washing machines, clothes washer-driers and   SI 1148-29(1)
dishwashers *shall* be economical in the use of water.

Note: This requirement *shall* be deemed to be satisfied in the case of machines having water consumption per cycle of not greater than the following:

- for domestic horizontal axis washing machines, 27 litres per kilogram of washload for a standard 60°C cotton cycle;
- for domestic washer-driers, 48 litres per kilogram of washload for a standard 60°C cotton cycle;
- for domestic dishwashers, 4.5 litres per place setting.

2.15.10 Water intended for outside use

Every pipe which conveys water to a drinking vessel for   SI 1148-30
animals or poultry *shall* be fitted with:

- a float-operated valve, or some other no less effective device to control the inflow of water, which:
  - is protected from damage and contamination; and
  - prevents contamination of the water supply; and
- a stopvalve or servicing valve as appropriate.

Every pond, fountain or pool *shall* have an impervious   SI 1148-31
lining or membrane to prevent the leakage or seepage of water.

2.15.11 Testing installations

Every water system *shall* be tested, flushed and where necessary disinfected before it is first used (Table 2.4).

Table 2.4 Tests for water systems that do not include plastic pipes

Test A	Test B
(i) The whole system is subjected to the test pressure by pumping for 30 min, after which the test continues for 90 min without further pumping;	(i) The whole system is subjected to the test pressure by pumping for 30 min, after which the pressure is noted and the test continues for 150 min without further pumping;
(ii) the pressure is reduced to one-third of the test pressure after 30 min;	(ii) the drop in pressure is less than 0.6 bar (60 kPa) after the following 30 min, or 0.8 bar (80 kPa) after the following 150 min; and
(iii) the pressure does not drop below one-third of the test pressure over the following 90 min; and	(iii) there is no visible leakage throughout the test
(iv) there is no visible leakage throughout the test	

3

What are the Building Regulations?

By Act of Parliament, councils are responsible for ensuring that the health, welfare and convenience of persons living and/or working in (or nearby) buildings are secured. This is achieved via Building Acts and Regulations as shown in Table 3.1.

Table 3.1 **Building legislation**

	Act	Regulations	Implementation
England and Wales	Building Act 1984	Building Regulations 2000	Approved Documents
Scotland	Building (Scotland) Act 2003	Building (Scotland) Regulations 2004	Technical Handbooks
Northern Ireland	Building Regulations (Northern Ireland) Order 1979	Building Regulations (Northern Ireland) 2000	'Deemed to satisfy' by meeting supporting publications

3.1 Building legislation

The prime purpose of building legislation is to prevent waste, undue consumption, misuse and contamination of water and to assist in the conservation of fuel and power. It imposes on owners and occupiers of buildings a set of requirements concerning the design and construction of buildings and the provision of services, fixtures and fittings used in (or in connection with) buildings. These requirements involve, and cover:

- a method of controlling (inspecting and reporting) buildings and their services, fittings and equipment;
- how services, fittings and equipment may be used;
- the inspection and maintenance of services, fittings or equipment used.

3.1.1 What happens if I contravene any of these requirements?

If you contravene the building legislation or wilfully obstruct a person acting in the execution of one of the Building Acts and/or any of their associated

Building Regulations then, on summary conviction, you could be liable to a fine or, in exceptional circumstances, even a short holiday in one of HM Prisons!

3.1.2 What about civil liability?

It is an aim of the building legislation that all building work is completed safely and without risk to people employed on the site or visiting the site, etc. Any contravention of the building legislation that causes injury (or death) to any person is liable to prosecution in the normal way.

3.1.3 Who polices the Act?

Under the terms of the current building legislation, local authorities are responsible for ensuring that any building work (e.g. construction and mainte-nance of sewers and/or drains and the laying and maintenance of water mains and pipes) is completed in conformance to the requirements of the associated Building Regulations.

The local authorities have the authority to:

- make the owner or occupier of any premises complete essential and reme-dial work in connection with the building legislation (particularly with respect to the construction, laying, alteration or repair of a sewer or drain);
- make you take down and remove or rebuild anything that contravenes a regulation;
- make you complete alterations so that your work complies with the Building Regulations;
- complete remedial and essential work themselves (if the owner or occu-pier refuses to do this work himself) and/or employ a third party to take down and rebuild non-conforming buildings or parts of buildings, and then send you the bill!

They can, in certain circumstances, even take you to court and have you fined, especially if you fail to complete the removal or rebuilding of the non-con-forming work.

The above authority to prosecute and order remedial work to be completed applies equally whether you are the actual owner or merely the occupier – so be warned!

3.2 Building Regulations

From the point of view of water systems, Building Regulations primarily include water services, fixtures, fittings and equipment, but also cover:

- cesspools (and other methods for treating and disposing of foul matter);
- drainage (including waste disposal units);
- materials and components (suitability, durability and use);

- prevention of infestation;
- resistance to moisture and decay;
- waste (storage, treatment and removal);
- wells and boreholes for supplying water;
- electrical safety;
- emission of gases, fumes, or other noxious and/or offensive substances;
- fire precautions.

and matters connected with (or ancillary to) any of the above matters.

3.2.1 What is the purpose of the Building Regulations?

The Building Regulations are legal requirements laid down by parliament, based on the Building Act 1984. They are approved by parliament and deal with the minimum standards of design and building work for the construction of domestic, commercial and industrial buildings.

Building Regulations ensure that new developments or alterations and/or extensions to buildings are all carried out to an agreed standard that protects the health and safety of people in and around the building.

Building standards are enforced by the local building control officer, but for matters concerning drainage or sanitary installations, you will need to consult their technical services department.

Builders and developers are required by law to obtain Building Control approval, which is an independent check that the Building Regulations have been complied with. There are two types of Building Control providers: the local authority and approved inspectors.

3.2.2 What building work is covered by the Building Regulations?

The Building Regulations cover ALL new building work.

This means that if you want to put up a new building, extend or alter an existing one, or provide new and/or additional fittings in a building such as drains or heat-producing appliances, washing and sanitary facilities and hot water storage (particularly unvented hot water systems), the Building Regulations will probably apply. Statutory Instrument 2006 No. 652 (SI 652) has amended the Building Regulations to cover situations where a building becomes a building to which energy efficiency requirements would now apply.

3.2.3 Are there any exemptions from Building Regulations?

The following are exempt from the Building Regulations:

- buildings belonging to 'statutory undertakers' (e.g. a water board);
- a 'public body' (i.e. local authorities, county councils and any other body 'that acts under an enactment for public purposes and not for its own profit').

This can be rather a grey area and it is best to seek advice if you think that you come under this category.

3.2.4 What about planning permission and Building Regulations approval?

Before undertaking any building project, you must first obtain the approval of local government authorities. There are two main controls that districts rely on to ensure that adherence to the local plan is ensured, namely planning permission and Building Regulation approval.

Although both of these controls are associated with gaining planning permission, actually receiving planning permission does not automatically confer Building Regulation approval and vice versa. You *may* require *both* before you can proceed. Indeed, there may be variations in the planning requirements, and to some extent the Building Regulations, from one area of the country to another.

Provided, however, that the work you are completing does not affect the external appearance of the building, you are allowed to make certain changes to your home without having to apply to the local council for permission. These are called permitted development rights, but the majority of building work that you are likely to complete will still require you to have planning permission – so be warned!

The actual details of planning requirements are complex, but for most domestic developments the planning authority is only really concerned with construction work such as an extension to the house or the provision of a new garage or new outbuildings that is being carried out. Structures such as walls and fences also need to be considered because their height or siting may infringe the rights of neighbours and other members of the community. The planning authority will also want to approve any change of use, such as converting a house into flats or running a business from premises previously occupied as a dwelling only.

Planning consent *may* be needed for minor works such as television satellite dishes, dormer windows, construction of a new access, fences, walls and garden extensions. You are advised to consult with Development Control staff before going ahead with such minor works.

Even when planning permission is not required, most building works, including alterations to existing structures, are subject to minimum standards of construction to safeguard public health and safety.

3.2.5 How is my building work evaluated for conformance with the Building Regulations?

Part of the local authority's duty is to make regular checks that all building work being completed is in conformance with the approved plan and the relevant building legislation. These checks would normally be completed at certain stages of the work (e.g. the excavation of foundations) and tests will include:

- tests of the soil or subsoil of the site of the building;
- tests of any material, component or combination of components that has been, is being, or is proposed to be used in the construction of a building;
- tests of any service, fitting or equipment that has been, is being, or is proposed to be provided in or in connection with a building.

The cost of carrying out these tests will normally be charged to the owner or occupier of the building and the local authority has the power to ask the person responsible for the building work to complete some of these tests on their behalf.

3.3 What about water supplies?

The Building Act stipulates that plans for proposed buildings will ensure that all occupants of the house will be provided with a supply of 'wholesome water, sufficient for their domestic purposes'. This can be achieved by:

- connecting the house to water supplies from the local water authority (normally referred to as the 'statutory water undertaker');
- otherwise taking water into the house by means of a pipe (e.g. from a local recognized supply);
- providing a supply of water within a reasonable distance from the house (e.g. such as from a well).

If an occupied house is not within a reasonable distance of a supply of 'wholesome water' or if the local authority is not satisfied that the water supply is capable of supplying 'wholesome water', then they can give notice that the owner of the building must provide water within a specified time. They also have the authority to prohibit the building from being occupied.

3.3.1 What happens if there is more than one property?

Where the local authorities are satisfied that two or more houses can most conveniently be met by means of a joint supply, they may give notice accordingly.

3.3.2 Can I ask the local authority to provide me with a supply of water?

If you are unable to provide a suitable supply of water, the local authority can provide, or secure the provision of, a supply of water to the house or houses in question and then recover any expenses reasonably incurred from the owner of the house or (where two or more houses are concerned) the owners of those houses.

The maximum amount that a local authority can charge for providing a suitable supply of water is £3000 in respect of any one house.

Where a supply of water is provided to a house by statutory water undertakers, water rates will be included in the normal rateable value of the house.

Where two or more houses are supplied with water by a common pipe belonging to the owners or occupiers of those houses, the local authority may, when necessary, repair or renew the pipe and recover any expenses reasonably incurred by them from the owners or occupiers of the houses.

3.3.3 What about plumbing?

Although planning permission is not required for plumbing replacements, it would be wise to consult the technical services department for any installation that alters present internal or external drainage. Building Regulation approval is, however, required for the installation or replacement of any hot water system if the water heater is unvented (i.e. supplied directly from the mains without an open expansion tank and with no vent pipe to atmosphere) and has storage capacity greater than 15 litres.

Many years ago the demand for external tanks for capturing rainwater made their installation quite commonplace. But it is rare today to need extra storage tanks, unless you are in a rural position. If you are considering installing an external water tank, then you should seek guidance from the local authority, especially if the tank is to be mounted on a roof.

3.3.4 What happens if the plans mean building over an existing sewer, etc.?

Before the local authority can approve a plan for building work which means having first to erect a building or an extension over an existing sewer or drain, they must notify and seek the advice of the water authority.

As part of the Public Health Act 1936 and the Control of Pollution Act 1974, local authorities are required to keep maps of all sewers, etc.

3.3.5 What about the buildings and drainage to buildings in Inner London?

Under the terms of the Building Act 1984, it is not lawful in an Inner London borough to erect a house/other building, or to rebuild a house/other building that has been pulled down to (or below) floor level, *unless* that house/building is provided with drains in conformance with the borough council's requirements. These drains must be suitable for the drainage of the *whole* building and all works, apparatus and materials used in connection with these drains must satisfy the council's requirements.

It is unlawful to occupy a house or other building in Inner London that has been erected or rebuilt in contravention of the above restriction.

The basic requirements of all Inner London borough councils are that:

- drains must be connected into a sewer that is (or is intended to be constructed) nearby;
- if a suitable sewer is not available, then a covered cesspool or other place should be used, *provided* that it is not under any house or other building;
- the drains must provide efficient gravitational drainage at all times and under all circumstances and conditions.

If it is impossible or unfeasible to provide gravitational drainage to all parts of the building, then (but depending on the circumstances) the council 'may' allow pumping and/or some other form of lifting apparatus to be used.

In *all* circumstances the council has the authority to order the owner/occupier:

- to construct a covered drain from the house or building into the sewer;
- to provide proper paved or water-resistant sloping surfaces for carrying surface water into the drain;
- to provide proper sinks, inlets and outlets (siphoned or otherwise trapped) for preventing the emission of effluvia from the drain, or any connection to it;
- to provide a proper water supply and water-supplying pipes, cisterns and apparatus for scouring the drain;
- to provide proper sand traps, expanding inlets and other apparatus for preventing the entry of improper substances into the drain.

If a house or building in an Inner London borough (regardless of when it was first erected), has insufficient drainage and there is no proper sewer within 200 feet of any part of the house or building, the Borough Council may serve on the owner written notice requiring that person:

- to construct a covered watertight cesspool or tank or other suitable receptacle (provided that it is not under the house); and
- to construct and lay a covered drain leading from the house or building into that cesspool, tank or receptacle.

The Inner London borough councils have the authority to carry out irregular inspections of drains and cesspools constructed by the owner and, if they prove to be unsuitable, they have the authority to make the owner alter, repair or abandon them if they contravene council regulations.

You are not allowed to commence any work on drains, dig out the foundations of a house or to rebuild a house in Inner London unless, at least seven days previously, you have provided a notice of intent to the Borough Council.

3.4 What are approved documents and technical handbooks?

These are a series of documents which are intended to provide practical guidance with respect to the requirements of the Building Regulations.

As previously mentioned, within England and Wales the Building Act imposes on owners and occupiers of buildings a set of requirements concerning the design and construction of buildings and the provision of services, fittings and equipment used in (or in connection with) buildings. These involve, and cover:

- a method of controlling (inspecting and reporting) buildings;
- how services, fittings and equipment may be used;
- the inspection and maintenance of any service, fitting or equipment used.

Details of these requirements are available in a series of documents called Approved Documents, which are intended to provide practical guidance with respect to the requirements of the Building Regulations.

Within Scotland, the requirements for buildings are controlled by the Building (Scotland) Act 2003, and the Building (Scotland) Regulations 2004 set the functional standards under this Act. The methods for implementing these requirements are similar to England and Wales, except that the guidance documents (i.e. for achieving compliance) are contained in two Technical Handbooks, one for domestic work and one for non-domestic. Each handbook has a general section and six technical sections.

The Building Regulations (Northern Ireland) Order 1979 (as amended by the Planning and Building Regulations (Amendment) (NI) Order 1990) is the main legislation for Northern Ireland, and the Building Regulations (Northern Ireland) 2000 detail the requirements for meeting this legislation.

Supporting publications (such as British Standards (BS), Building Research Establishment (BRE) publications and/or Technical Booklets published by the Department) are used to ensure that the requirements are implemented (*deemed to satisfy*).

Table 3.2 provides a comparison between these different (but very similar) pieces of legislation.

3.5 How do the requirements of the Building Regulations affect water services?

Even when planning permission is not required, most building works, including alterations to existing structures, are subject to minimum standards of construction to safeguard public health and safety. The Building Regulations cover all new building work and this means that if you want to provide new

Table 3.2 Approved Documents, Technical Handbooks and Technical Booklets

England and Wales		Scotland		Northern Ireland	
Part A	Structure	Section 1	Structure	Technical Booklet D	Structure
Part B	Fire safety	Section 2	Fire	Technical Booklet E	Fire safety
Part C	Site preparation and resistance to contaminants and water	Section 3	Environment	Technical Booklet C	Preparation of site and resistance to moisture
Part D	Toxic substances	Section 3	Environment	Technical Booklet B	Materials and workmanship
Part E	Resistance to the passage of sound	Section 5	Noise	Technical Booklet G	Sound insulation of dwellings
Part F	Ventilation	Section 3	Environment	Technical Booklet K	Ventilation
Part G	Hygiene	Section 3	Environment	Technical Booklet P	Sanitary appliances and unvented hot water storage systems
Part H	Drainage and waste disposal	Section 3	Environment	Technical Booklet J / Technical Booklet N	Solid waste in buildings / Drainage
Part J	Combustion appliances and fuel storage systems	Section 3	Environment	Technical Booklet L	Heat-producing appliances and liquefied petroleum gas installations
Part K	Protection from falling, collision and impact	Section 4 / Section 4	Safety / Safety	Technical Booklet H	Stairs, ramps and protection from impact
Part L	Conservation of fuel and power	Section 6	Energy	Technical Booklet F	Conservation of fuel and power
Part M	Access and facilities for disabled people	Section 4	Safety	Technical Booklet R	Access for facilities and disabled people
Part N	Glazing	Section 6	Energy	Technical Booklet V	Glazing
Part P	Electrical safety	Section 4	Safety		

and/or additional fittings in a new building, extension or alteration such as drains or heat-producing appliances, washing and sanitary facilities and hot water storage (particularly unvented hot water systems), the Building Regulations will probably apply.

 Note: The mandatory requirements of the Regulations concerning water are shown below.

3.5.1 Site preparation

- Before any building works commence, all vegetation and topsoil are removed;
- contaminated ground shall be either treated, neutralized or removed before a building is erected;
- subsoil drainage shall be provided to waterlogged sites.

3.5.2 Resistance to moisture

The floors, walls and roof of the building:

- should not be adversely affected by interstitial condensation;
- shall protect the building and people who use the building from harmful effects caused by the spillage of water from (or associated with) sanitary fittings and/or fixed appliances.

3.5.3 Protection against sound within a dwelling-house, etc.

Dwelling-houses, flats and rooms for residential purposes shall be designed and constructed so that the internal walls between a bedroom or a room containing a water closet, and other rooms and internal floors shall provide reasonable resistance to sound.

3.5.4 Sanitary conveniences and washing facilities

ALL buildings are required to have satisfactory sanitary conveniences and washing facilities and *ALL* dwellings (house, flat or maisonette) should have at least *one* closet and *one* washbasin which should be separated by a door from any space used for food preparation or where washing-up is done. They shall be provided so that:

- the surfaces of a closet, urinal or washbasin are smooth, non-absorbent and capable of being easily cleaned;
- closets (and/or urinals) are capable of being flushed effectively;
- closets (and/or urinals) are only connected to a flush pipe or discharge pipe;

- closets fitted with flushing apparatus should discharge through a trap and discharge pipe into a discharge stack or a drain;
- washbasins are, ideally, located in the room containing the closet;
- washbasins have a supply of hot and cold water.

3.5.5 Bathrooms

All dwellings (house, flat or maisonette) should have *at least* one bathroom with a fixed bath or shower and the bath or shower should:

- have a supply of hot and cold water;
- discharge through a grating, a trap and branch discharge pipe to a discharge stack or (if on a ground floor);
- discharge into a gully or directly to a foul drain;
- be connected to a macerator and pump (of an approved type) if there is no suitable water supply or means of disposing of foul water.

3.5.6 Drainage and waste disposal

- New drains taking foul water from buildings are required to discharge into a foul water sewer (or other suitable outfall), be watertight and be accessible for cleaning;
- where no public sewer is available, holding tanks or sewage treatment plants should be made available;
- new drains taking rainwater from roofs of buildings need to be watertight, accessible for cleaning and (if there is no sewer available) discharge to a suitable surface water sewer or ditch, soakaway or watercourse;
- storage facilities, reasonably close to the building, need to be provided for refuse collection.

3.5.7 Rainwater drainage

Rainwater drainage systems shall:

- minimize the risk of blockage or leakage;
- be accessible for clearing blockages;
- ensure that rainwater soaking into the ground is distributed sufficiently so that it does not damage foundations of the proposed building or any adjacent structure;
- ensure that rainwater from roofs and paved areas is carried away from the surface either by a drainage system or by other means;
- ensure that the rainwater drainage system carries the flow of rainwater from the roof to an outfall (e.g. a soakaway, a watercourse, surface water or a combined sewer).

3.5.8 Foul water drainage

The foul water drainage system shall:

- convey the flow of foul water to a foul water outfall (i.e. sewer, cesspool, septic tank or settlement (i.e. holding) tank);
- minimize the risk of blockage or leakage;
- prevent foul air from the drainage system from entering the building under working conditions;
- be ventilated;
- be accessible for clearing blockages;
- not increase the vulnerability of the building to flooding.

3.5.9 Separate systems of drainage

Separate systems of drains and sewers shall be provided for foul water and rainwater where the rainwater is not contaminated; and the drainage is to be connected either directly or indirectly to the public sewer system and either:

- the public sewer system in the area shall comprise separate systems for foul water and surface water; or
- a system of sewers is provided for the separate conveyance of surface water either by the sewerage undertaker or by some other person (where the sewer is the subject of an agreement to make a declaration of vesting pursuant to Section 104 of the Water Industry Act 1991).

3.5.10 Wastewater treatment systems and cesspools

Wastewater treatment systems shall:

- have sufficient capacity to enable breakdown and settlement of solid matter in the wastewater from the buildings;
- be sited and constructed so as to prevent overloading of the receiving water.

Cesspools shall have sufficient capacity to store the foul water from the building until they are emptied.

Wastewater treatment systems and cesspools shall be sited and constructed so as not to:

- be prejudicial to health or a nuisance;
- adversely affect water sources or resources;
- pollute controlled waters;
- be in an area where there is a risk of flooding.

Septic tanks and wastewater treatment systems and cesspools shall be constructed and sited so as to:

- have adequate ventilation;
- prevent leakage of the contents and ingress of subsoil water (having regard to water table levels at any time of the year and rising groundwater levels).
  - drainage fields shall be sited and constructed so as to:
  - avoid overloading of the soakage capacity; and
  - provide adequately for the availability of an aerated layer in the soil at all times.

3.5.11 Building over sewers

Building, extension or work involving underpinning shall:

- be constructed or carried out in a manner that will not overload or otherwise cause damage to the drain, sewer or disposal main either during or after the construction;
- not obstruct reasonable access to any manhole or inspection chamber on the drain, sewer or disposal main;
- reduce the risk of damage to the building as a result of failure of the drain, sewer or disposal main.

 Note: In the event of the drain, sewer or disposal main requiring replacement, it should not unduly obstruct work to replace the drain, sewer or disposal main, on its present alignment.

3.5.12 Hot water storage

Unvented hot water systems over a certain size are required to have safety provisions to prevent explosion and shall:

- be installed by a competent person;
- not exceed 100°C;
- discharge safely;
- not cause danger to persons in or about the building.

3.5.13 Conservation of fuel and power

Reasonable provision shall be made for the conservation of fuel and power in buildings by limiting heat gains and losses through thermal elements and other parts of the building fabric and from pipes, ducts and vessels used for space heating, space cooling and hot water services.

3.5.14 Performance tests

Energy efficiency measures shall be provided which provide information, in a suitably concise and understandable form (including results of performance tests carried out during the works) that shows building occupiers how the heating and hot water services can be operated and maintained.

3.5.15 Electrical earthing

- Main equipotential bonding conductors are required for *all* water service pipes.
- The installation of supplementary equipotential bonding conductors is required for installations and locations where there is an increased risk of electric shock (e.g. such as bathrooms and shower rooms).

Note: The most usual type of earthing is an electricity distributor's earthing terminal, which is provided for this purpose, near the electricity meter.

It is NOT permitted to use a water pipe as a means of earthing for an electrical installation (this does not rule out, however, equipotential bonding conductors being connected to these pipes).

For further information concerning Building Regulations, the reader may like to look at one of my other 'In Brief' books (Table 3.3).

Table 3.3 'In Brief' books

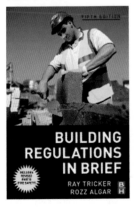

Building Regulations in Brief (5th edition) Butterworth-Heinemann ISBN-13: 978-0-7506-8444-6	This handy and affordable guide is a time-saver for both professionals and enthusiasts. The information is sensibly organized by building element rather than by regulation, so that you can quickly lay your hands on whatever you need to know from whichever document. The benefits and requirements of each regulation are clearly explained, as are history, current status, associated documentation and how local authorities and council view their importance. This new edition includes: • The new Regulatory Reform (Fire Safety) Order and what this means for Part B (Fire Safety) • Updates to Part L (Energy Efficiency) • An improved user-friendly index • Annexes covering: Access and facilities for disabled people; Conservation of fuel and power; Sound insulation and Electrical Safety provided online

(Continued)

Table 3.3 Continued

Scottish Building Standards in Brief Butterworth-Heinemann ISBN-13: 978-0-7506-8558-0 	*Scottish Building Standards in Brief* is specifically for the Scottish Building Standards and is an ideal book for builders, architects, designers and DIY enthusiasts working in Scotland. The meaning of the regulations, their history, current status, requirements, associated documentation and how local authorities view their importance are fully covered. The book emphasizes the benefits and requirements of each one and these are explained in a user-friendly manner There is no easier or clearer guide to help you to comply with the Scottish Building Standards in the simplest and most cost-effective manner possible
Wiring Regulations in Brief Butterworth-Heinemann ISBN-13: 978-0-7506-8973-0 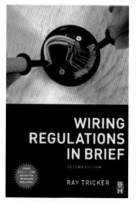	This handy guide provides an on-the-job reference source for electricians, designers, service engineers, inspectors, builders, students and DIY enthusiasts. Topic-based chapters link areas of working practice, such as cables, installations, testing and inspection and special locations with the specifics of the Regulations. This allows quick and easy identification of the official requirements relating to the situation in front of you. Packed with useful hints and tips, and highlighting the most important or mandatory requirements Note: This book is a concise reference on *all* aspects of the 17th edition of the *IEE Wiring Regulations*.

Part II

4

Meeting the Requirements of the Building Regulations

As noted previously, Approved Documents (Technical Handbooks and Technical Booklets) are a series of documents that are intended to provide practical guidance with respect to the requirements of the Building Regulations and the Water Regulations.

4.1 Compliance

There is no obligation to adopt any particular solution that is contained in any of these guidance documents, especially if you prefer to meet the relevant requirement in some other way. However, should a contravention of a requirement be alleged, if you have followed the guidance in the relevant Approved Documents (Technical Handbooks and/or Booklets) that will be evidence tending to show that you have complied with the Regulations. If you have not followed the guidance, then that will be seen as evidence tending to show that you have not complied with the requirements and it will then be up to you, the builder, architect and/or client, to demonstrate that you have satisfied the requirements of the Building Regulations.

This compliance may be shown in a number of ways such as using:

- a product bearing CE marking (in accordance with the Construction Products Directive (89/106/EEC) as amended by the CE Marking Directive (93/68/EEC) as implemented by the Construction Products Directive 1994 (SI 1994/3051));
- an appropriate technical specification (as defined in the Construction Products Directive – 89/1 06/EEC);
- a recognized British Standard;
- a British Board of Agreement Certificate;
- an alternative, equivalent national technical specification from any member state of the European economic area, or Turkey;

- a product covered by a national or European certificate issued by a European Technical Approval issuing body;

and by demonstrating professional workmanship.

4.1.1 Materials and workmanship

As stated in the various Building Acts, 'Any building work which is subject to Building Regulations should be carried out with proper materials and in a workmanlike manner'.

4.1.2 What materials can I use?

Other than the two exceptions below, provided that the materials and components you have chosen to use are from an approved source and are of approved quality (CE marking in accordance with the Construction Products Directive (89/106/EEC), the Low Voltage Directive (73/23/EEC and amendment 93/68 EEC) and the EMC Directive (89/336/EEC) as amended by the CE Marking Directive (93/68/EEC)) then the choice is fairly unlimited.

Short-lived materials

Even if a plan for building work complies with the Building Regulations, if this work has been completed using short-lived materials (i.e. materials that are, in the absence of special care, liable to rapid deterioration) the local authority can:

- reject the plans;
- pass the plans subject to a limited-use clause (on expiration of which, that material will have to be removed);
- restrict the use of the building.

Unsuitable materials

If, once building work has begun, it is discovered that it has been made using materials or components that have been identified by the Secretary of State (or his nominated deputy) as being unsuitable materials, the local authority has the power to:

- reject the plans;
- fix a period in which the offending work must be removed;
- restrict the use of the building.

If the person completing the building work fails to remove the unsuitable material or component(s), then that person is liable to be prosecuted and, on summary conviction, faces a heavy fine.

4.1.3 Technical specifications

In addition to what one would normally expect, Building Regulations may be made for specific purposes such as:

- health and safety;
- welfare and convenience of disabled people;
- conservation of fuel and power;
- prevention of waste or contamination of water.

These are aimed at furthering the protection of the environment and facilitating sustainable development.

Although the main requirements for health and safety are now covered by the Building Regulations, there are still some requirements contained in the Workplace (Health, Safety and Welfare) Regulations 1992 that may need to be considered as they could contain requirements that affect building design.

Note: For further information see Workplace (Health, Safety and Welfare) Regulations 1992. Approved Code of Practice L24, published by HSE Books 1992 (ISBN 0-7176-0413-6).

Standards and technical approvals, as well as providing guidance, address other aspects of performance such as serviceability and/or other aspects related to health and safety not covered by the Regulations.

When an Approved Document makes reference to a named standard, the relevant version of the standard is the one listed at the end of that particular Approved Document. However, if this version of the standard has been revised or updated by the issuing standards body, the new version may be used as a source of guidance provided it continues to address the relevant requirements of the Regulations.

4.1.4 Independent certification schemes

Within the UK there are many product certification schemes that certify compliance with the requirements of a recognized standard or document that is suitable for the purpose and material being used. Certification bodies that approve such schemes will normally be accredited by the United Kingdom Accreditation Service (UKAS).

4.1.5 Standards and technical approvals

Standards and technical approvals provide guidance related to the Building Regulations and address other aspects of performance such as serviceability or aspects which, although they relate to health and safety, are not covered by the Regulations.

4.1.6 European pre-standards

The British Standards Institution (BSI) will be issuing Pre-standard (ENV) Structural Eurocodes as they become available from the European Standards Organization, Comité Européen de Normalization (CEN) and Comité Européen de Normalization Electrotechnique (CENELEC).

For example, DD ENV 1992-1-1: 1992 Eurocode 2: Part 1 and DD ENV 1993-1-1: 1992 Eurocode 3: Part 1-1 *General Rules* and *Rules for Buildings in Concrete and Steel* have been thoroughly examined over a period of several years and are considered to provide appropriate guidance when used in conjunction with their national application documents for the design of concrete and steel buildings, respectively.

When other ENV Eurocodes have been subjected to a similar level of examination they may offer an alternative approach to Building Regulation compliance and, when they are eventually converted into fully approved European Normalization (EN) standards, they will be included as referenced standards in the guidance documents.

Note: If a national standard is going to be replaced by a European harmonized standard, then there will be a coexistence period during which either standard may be referred to. At the end of the coexistence period the national standard will be withdrawn.

When reading the following chapters, you will probably notice that quite a number of the requirements are shown in more than one chapter, rather than showing a requirement only once and then making a cross-reference to it. This has been done deliberately to save the reader having constantly to turn backwards and forwards to find the relevant page.

5

Design and Installation

Water systems and fittings in premises that are, or will be, connected to the public water supply must comply with the Water Regulations *and* the Building Regulations, and most plumbing work such as the installation of water fittings in connection with the:

- erection of any new building or structure;
- extension or alteration of the water system in any premises except a domestic dwelling;
- material change in use of any premises;
- construction of a large pond or swimming pool with automatic replenishment;
- installation of any fitting listed in section 5 of the Regulations;

will *also* need the prior consent of the water company before you can commence work.

The two main areas that have to be considered when designing and constructing any dwelling or building are:

- the requirements of the Water Regulations;
- the requirements of the Building Regulations (Figure 5.1) and in particular:
  - the environment and its effect on the comfort of occupants and its sustainability;
  - ventilation and its effect on the health of occupants;
  - energy and the need to reduce carbon dioxide emissions;
  - safety and the need to safeguard occupants from inherent risks and danger;
  - commissioning and the need to ensure the safe operation of all systems.

The following are extracts from the Water Regulations/Bye-laws, the UK's Building Regulations and (where appropriate) the Wiring Regulations that directly affect the supply and use of water in buildings and dwellings and need to be considered by both the designer and the installer.

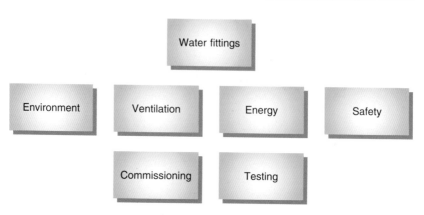

Figure 5.1 Building Regulation requirements concerning the design and construction of water fittings

5.1 What are the mandatory requirements for water fittings?

No material or substance, either alone or in combination with any other material or substance, or with the contents of any water fitting of which it forms a part, which causes or is likely to cause contamination of water shall be used in the construction, installation, renewal, repair or replacement of any water fitting which conveys or receives, or may convey or receive, water supplied for domestic or food production purposes.

This requirement does not apply to a water fitting that is downstream of a terminal fitting that supplies wholesome water where:

- the use to which the water downstream is put does not require wholesome water; and
- a suitable arrangement or device to prevent backflow has been installed.

5.2 What are the mandatory requirements from the Building Regulations?

- *All occupants of a house will be provided with a supply of 'wholesome water, sufficient for their domestic purposes' by:*
  - *connecting the house to water supplies from the local water authority;*
  - *otherwise taking water into the house by means of a pipe (e.g. from a local recognized supply); or*
  - *providing a supply of water within a reasonable distance from the house (e.g. from a well).*

- *All buildings are required to have satisfactory sanitary conveniences and washing facilities.*
- *All dwellings (house, flat or maisonette) should have at least one closet and one washbasin which should be separated by a door from any space used for food preparation or where washing-up is done.*
- *All dwellings (house, flat or maisonette) should have at least one bath-room with a fixed bath or shower.*
- *Special attention shall be taken to rainwater, wastewater and foul water drainage systems.*
- *Cesspools and sewers shall be properly constructed and regularly inspected and maintained.*
- *Unvented hot water systems over a certain size are required to have safety provisions to prevent explosion.*

In all Inner London borough councils:

- drains must be connected into a sewer that is (or is intended to be constructed) nearby;
- if a suitable sewer is not available, then a covered cesspool or other place should be used, provided that it is not under any house or other building;
- the drains must provide efficient gravitational drainage at all times and under all circumstances and conditions.

5.3 Water fittings

Every water fitting *shall*:

be immune to (or protected from) corrosion by galvanic action or by any other process which is likely to result in contamination or waste of water;	SI 1148-3(a)
be constructed of materials of such strength and thickness as to resist damage from any external load, vibration, stress or settlement, pressure surges, or temperature fluctuation to which it is likely to be subjected;	SI 1148-3(b)
be watertight;	SI 1148-4(a)
be constructed and installed so as to:	SI 1148-4(b)
o prevent ingress by contaminants; and	
o inhibit damage by freezing or any other cause;	
be installed so as to minimize the risk of permeation by, or deterioration from contact with, any substance which may cause contamination;	SI 1148-4(c)
be adequately supported;	SI 1148-4(d)

- be capable of withstanding an internal water   SI 1148-5
  pressure not less than 1½ times the maximum
  pressure to which that fitting has been designed to
  be subjected whilst in operation.

No water fitting *shall*:

- be installed, connected or used which is likely   SI 1148-6
  to have a detrimental effect on the quality or
  pressure of water in a water main or other pipe
  belonging to a water undertaker;
- be embedded in any wall or solid floor.   SI 1148-7(1)

No fitting which is designed to be operated or   SI 1148-7(2)
maintained, whether manually or electronically, or which
consists of a joint, *shall* be a concealed water fitting.

Any concealed water fitting or mechanical backflow   SI 1148-7(3)
prevention device, not being a terminal fitting, *shall*
be made of gunmetal, or another material resistant to
dezincification.

Any water fitting laid below ground level *shall* be   SI 1148-7(4)
sufficiently deep enough to prevent water freezing in the
fitting

Note: 'Concealed water fitting' means a water fitting which:
- is installed below ground;
- passes through or under any wall, footing or foundation;
- is enclosed in any chase or duct; or
- is in any other position which is inaccessible or renders access difficult.

No water fitting shall be installed in such a position, or pass through such surroundings, that it is likely to cause contamination or damage to the material
of the fitting or the contamination of water supplied by the water undertaker.

Any pipe supplying cold water for domestic purposes to SI 1148-9
any tap *shall* be so installed that (as far as is reasonably
practicable) the water is not warmed above 25°C.

Every supply pipe or distributing pipe that is intended   SI 1148-10(1)
to supply water to separate premises *shall* be fitted with
a stopvalve that is conveniently located so as to enable
the supply to those premises to be shut off without
shutting off the supply to any other premises.

Where a supply pipe or distributing pipe provides water   SI 1148-10(2)
in common to two or more premises, it *shall* be fitted
with a stopvalve which each occupier of those premises
has access to.

Water supply systems *shall* be capable of being   SI 1148-11
drained down and be fitted with an adequate number of
servicing valves and drain taps so as to minimize the
discharge of water when water fittings are maintained
or replaced.

A sufficient number of stopvalves *shall* be installed for
isolating parts of the pipework.

The water system *shall* be capable of withstanding   SI 1148-12(1)
an internal water pressure not less than 1½ times the
maximum pressure to which the installation or relevant
part is designed to be subjected in operation (this is
usually referred to as 'the test pressure').

This requirement *shall* be deemed to be satisfied: SI 1148-12(2)

- in the case of a water system that does not
  include a pipe made of plastics, where:
  - the whole system is subjected to the test pressure
    by pumping, after which the test continues for
    one hour without further pumping;
  - the pressure in the system is maintained for
    one hour; and
  - there is no visible leakage throughout the test;
- in all other cases, where either of the tests shown
  above is satisfied.

5.4 Building Regulations: environment

These days, when you hear people talk about 'the environment', they are
often referring to the overall condition of our planet, or how healthy it is.
From a building perspective, the environment is everything that makes up our
surroundings and affects our ability to live comfortably and in a sustainable
manner.

Buildings *must* be designed and constructed so that there will be no threat to the building or the health of the occupants as a result of flooding and the accumulation of groundwater.

SBR-D-3.3, SBR-ND-3.3

Buildings must *not* be constructed over an existing drain (including a field drain) that is to remain active.

SBR-D-3.5, SBR-ND-3.5

Every building, and hard surface within the curtilage of a building, *must* be designed and constructed with a surface water drainage system that will:

SBR-D-3.6, SBR-ND-3.6

- ensure the disposal of surface water without threatening the building and the health and safety of the people in and around the building; and
- have facilities for the separation and removal of silt, grit and pollutants.

Every private wastewater treatment plant or septic tank serving a building *must* be designed and constructed so that:

SBR-D-3.8, SBR-ND-3.8

- it *will* ensure the safe temporary storage and treatment of wastewater prior to discharge;
- the disposal of the wastewater to ground is safe and is not a threat to the health of the people in or around the building.

SBR-D-3.9, SBR-ND-3.9

Every wastewater drainage system serving a non-domestic building *must* be designed and constructed in such a way as to ensure the removal of wastewater from the building without threatening the health and safety of the people in and around the building.

SBR-ND-3.7

Every domestic building *must* be designed and constructed so that accommodation for solid waste storage is provided which does not contaminate any water supply, groundwater or surface water.

SBR-D-3.25

Every building *must* be designed and constructed so that an oil storage installation incorporating oil storage tanks (used solely to serve a fixed combustion appliance installation providing space heating or cooking facilities in a building) will:

SBR-D-3.24, SBR-ND-3.24

- reduce the risk of oil escaping from the installation;
- contain any oil spillage likely to contaminate any water supply, groundwater, watercourse, drain or sewer; and
- permit any spill to be disposed of safely.

Every non-domestic building *must* be provided with a water supply for use by the fire service.

SBR-D-2.13, SBR-ND-2.13

5.5 Building Regulations: ventilation

Ventilation of a building is required to prevent the accumulation of moisture that could lead to mould growth (and pollutants) originating from within the building that could become a risk to the health of the occupants.

Ventilation *must* have the capability of:

SBR-D-3.14.1, SBR-ND-3.14.1

- removing excess water vapour from areas where it is produced in significant quantities, such as kitchens, utility rooms, bathrooms and shower rooms, to reduce the likelihood of creating conditions that support the germination and growth of mould, harmful bacteria, pathogens and allergens;
- rapidly diluting water vapour, where necessary, that is produced in apartments and sanitary accommodation.

Ventilation *must* be to the outside (i.e. external) air.

SBR-D-3.14.1, SBR-ND-3.14.1

5.6 Building Regulations: energy

The current editions of the Building Regulations focus on the reduction of carbon dioxide emissions arising from the use of hot water heating and lighting in buildings and its guidance sets an overall level for maximum carbon dioxide emissions in buildings incorporating a range of parameters which will influence energy use.

This means that for all new buildings, *designers are now obliged* to consider energy as a complete package rather than only looking at individual elements such as insulation or boiler efficiency favouring, and localized or building-integrated low and zero carbon technologies (LZCT) (e.g. photovoltaics, active solar water heating, combined heat and power and heat pumps) can be used as a contribution towards meeting this standard.

> Every building *must* be designed and constructed so that: SBR-D-6.3, SBR-ND-6.3
>
> - the heating and hot water service systems installed are energy efficient and are capable of being controlled to achieve optimum energy efficiency;
> - temperature loss from heated pipes, ducts and vessels, and temperature gain to cooled pipes and ducts, is resisted;   SBR-D-6.4, SBR-ND-6.4
> - energy supply systems and building services which use fuel or power for heating the water are commissioned to achieve optimum energy efficiency.   SBR-D-6.7, SBR-ND-6.7

5.7 Building Regulations: safety

Safety has been defined by the International Standards Organization (ISO) as 'a state of freedom from unacceptable risks of personal harm'. This recognizes that no activity is absolutely safe or free from risk. Indeed, no building can be absolutely safe and some risk of harm to users may exist in every building. Building standards seek to limit risk to an acceptable level by identifying hazards in and around buildings, and the UK's Building Regulations provide recommendations for the design of buildings that will ensure access and usability and reduce the risk of accident. Amongst other areas, they are designed to locate safely hot water and steam vent pipe outlets, minimize the risk of explosion through malfunction of unvented hot water storage systems, prevent scalding by hot water from sanitary facilities, and ensure the appropriate location and construction of storage tanks for liquefied petroleum gas.

> Every building *must* be designed and constructed so that protection is provided for people in (and around) the building from the danger of severe burns or scalds from the discharge of steam or hot water.   SBR-D-4.9, SBR-ND-4.9
> Electrical wiring systems which are part of a water installation (or part of the building structure) *shall* be selected and erected so: WR-522.3.1
>
> - that no damage is caused by condensation or ingress of water during installation, use and/or maintenance;

- that damage arising from mechanical stress   WR-522.6.1
(e.g. by impact, abrasion, penetration, tension
or compression during installation, use or
maintenance) is minimized.

5.8 Building Regulations: commissioning building services

Where a water fitting is installed, altered, connected or disconnected by an approved contractor, the contractor *shall* upon completion of the work provide a signed certificate stating whether the water fitting complies with the requirements of these Building Regulations/Bye-laws to the person who commissioned the work,

and to the water undertaker if the work involves the construction of a pond or swimming pool with a capacity greater than 10,000 litres that is designed to be automatically replenished and filled with water supplied by a water undertaker.

A person who:

- contravenes the requirements for the installation of water fittings: or
- fails to provide a contractors certificate; or
- commences an operation listed in Table 5.1 without giving the necessary notice or consent;

is guilty of an offence and liable on summary conviction to a fine.

Commissioning (i.e. in terms of achieving the levels of energy efficiency that the component manufacturers expect from their product(s)) should also be carried out with a view to ensuring the safe operation of the water systems and surfaces.

Table 5.1 **Tests for water systems that do not include plastic pipes**

Test A	Test B
(i) The whole system is subjected to the test pressure by pumping for 30 min, after which the test continues for 90 min without further pumping;	(i) The whole system is subjected to the test pressure by pumping for 30 min, after which the pressure is noted and the test continues for 150 min without further pumping;
(ii) the pressure is reduced to one-third of the test pressure after 30 min;	(ii) the drop in pressure is less than 0.6 bar (60 kPa) after the following 30 min, or 0.8 bar (80 kPa) after the following 150 min; and
(iii) the pressure does not drop below one-third of the test pressure over the following 90 min; and	(iii) there is no visible leakage throughout the test
(iv) there is no visible leakage throughout the test	

All heating and hot water services should be inspected   SBR-D-6.7.1
and commissioned in accordance with manufacturers'
instructions to ensure optimum energy efficiency.

5.9 Testing installations

Every water system shall be tested, flushed and where necessary disinfected
before it is first used.

5.10 Written information

Correct use and maintenance of building services equipment is essential if the
benefits of enhanced energy efficiency are to be realized from such equipment. To achieve this it is essential that user and maintenance instructions
together with all other relevant documentation are available. Risk assessment
should be specific to each building site and take into account the presence of
source, pathways and receptors at a particular building site.

For a domestic building, written information   SBR-D-6.8.1
concerning the operation and maintenance of the
heating and hot water service systems (together with
any decentralized power generation equipment) should
be made available for the use of the occupier.

5.11 Location of an energy performance certificate

The Energy Performance Certificate should be indelibly   SBR-D-6.9.3
marked and located in a position that is readily accessible
(such as a cupboard containing the gas or electricity
meter or the water supply stopcock) and it should be
protected from weather and not easily obscured.

6

Site Preparation

There are two main types of building in common use today: those made of brick and those made of timber (Figure 6.1). There are many different styles of brick-built houses and, equally, there are various methods of construction.

Brickwork, as well as giving a building character, provides the main load-bearing element of the house and supports the weight of the structure. Most brick-built buildings are built on a solid base called a foundation.

Timber-framed houses are usually built on a concrete foundation with a 'strip' or 'raft' type construction to spread the weight. They differ from their brick-built counterparts in that the main structural elements are timber frames.

Although these two types of building are constructed differently, the requirements for the preparation on the site are very similar.

6.1 Requirements

The foundations of buildings should be safeguarded from the adverse effects of groundwater.

(BR-AD-C1)

The building *will* be constructed so that adequate subsoil drainage shall be provided if it is needed to avoid:

(a) the passage of ground moisture to the interior of the building;
(b) damage to the building, including damage through the transport of water-borne contaminants to the foundations of the building.

(BR-AD-C1)

Note: For the purpose of this requirement, 'contaminant' means any substance that is, or may become harmful to buildings, including substances that are corrosive, explosive, flammable, radioactive or toxic.

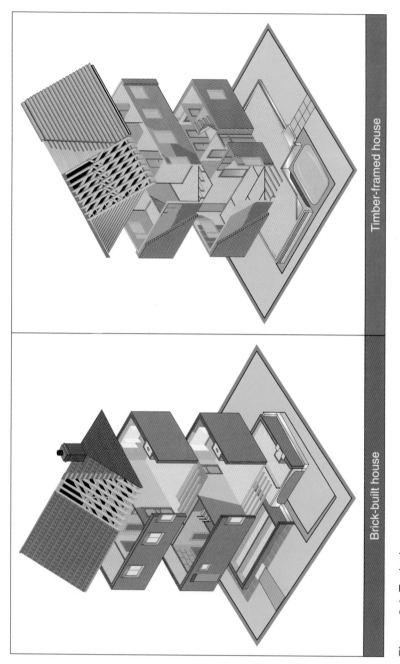

Timber-framed house

Brick-built house

Figure 6.1 Typical components

6.2 Meeting the requirements

In consideration to the potential risks to buildings and building materials as well as services, the following hazards shall be considered:

A detailed risk estimation and evaluation *shall* be   BR-AD-C2.13
carried out on the intended and surrounding ground,
particularly with respect to:
- floodwater affected by contaminants;   BR-AD-C2.23d
- waste matter or sewage;
- substances in the ground.

This is likely to involve collection and analysis of soil, soil gas, surface water and groundwater samples by the use of invasive and/or non-invasive techniques.

Surface soil and vegetable matter can be detrimental to a building's structure if left undisturbed within the building footprint.

To prevent water collecting under the building, the   SBR-D-3.1.1,
solum should, therefore, *not* be lower than the highest   SBR-ND-3.1.1
level of the adjoining ground.

6.2.1 Subsoil drainage

Where contaminants are present in the ground, consideration should be given to subsoil drainage to prevent the transportation of water-borne contaminants to the foundations or into the building or its services.

The selection of an appropriate drainage layout will depend on the nature of the subsoil and the topography of the ground.

Where the water table can rise to within 0.25 m of   BR-AD-C3.2
the lowest floor of the building, or where surface
water could enter or adversely affect the building,
either the ground to be covered by the building
should be drained by gravity, or other effective
means of safeguarding the building should be taken.

Cesspools and settlement tanks should prevent   BR-AP- H2 (1.63)
leakage of the contents and ingress of subsoil water.

6.2.2 Groundwater

Water is the prime cause of deterioration in building materials and constructions and the presence of moisture encourages mould growth which can
become a health hazard. Groundwater can penetrate building fabric from
below and will then rise vertically by capillary action.

The design and construction of new buildings needs to take into consideration the effect on existing buildings caused by changes in groundwater levels,
and the likely effects of flooding and measures to reduce the risk of flood
damage in dwellings should be taken into consideration by the developers.

In normal circumstances (i.e. unless you have the consent of the local authority) you are not allowed to construct a cellar or room in (or as part of) a
house, an existing cellar, a shop, inn, hotel or office if the floor level of the
cellar or room is lower than the ordinary level of the subsoil water on (under
or adjacent to) the site of the house, shop, inn, hotel or office.

Every building *must* be designed and constructed in such a way that there will not be a threat to the building, the occupants and/or the health of people in or around the building as a result of:	SBR-D-3.3, SBR-D-3.3
• flooding and the accumulation of groundwater;	
• moisture penetration from the ground;	SBR-D-3.4, SBR-ND-3.4
• rising damp which can either damage the building fabric or penetrate to the interior where it may constitute a health risk to occupants.	SBR-D-3.4.0, SBR-ND-3.4.0
All proposed building sites should be appraised initially to ascertain the risk of flooding of the land and an assessment made as to what effects the development may have on adjoining ground.	SBR-D-3.3.1, SBR-ND-3.3.1
Ground below and immediately adjoining a building that is liable to accumulate floodwater or groundwater requires treatment to be provided against the harmful effects of such water.	SBR-D-3.3.1, SBR-ND-3.3.1

Treatment could include a field drain system that is designed to:

• increase the stability of the ground;
• avoid surface flooding;
• alleviate subsoil water pressures likely to cause dampness to below-
  ground accommodation;
• assist in preventing damage to foundations of buildings;
• prevent frost heave of subsoil that could cause fractures to structures such
  as concrete slabs.

The selection of an appropriate drainage layout will depend on the nature of the subsoil and the topography of the ground.

6.2.3 Surface water

Climate change is expected to result in more rain in the future and it is essential that this is taken into account in today's buildings. Developers should also be aware of the dangers from possible surface water run-off from their building site to other properties and procedures should be in place to overcome this occurrence.

Every building, and hard surface within the curtilage of a building, *must* be designed and constructed with a surface water drainage system that will: • ensure the disposal of surface water without threatening the building and the health and safety of the people in and around the building; and • have facilities for the separation and removal of silt, grit and pollutants.	SBR-D-3.6, SBR-D-3.3.1
It is essential that surface water from buildings is removed quickly and safely without damage to the building or danger to people around the building, and does not pose a risk to the environment by flooding or pollution.	SBR-D-3.6.0, SBR-ND-3.6.0

The installation of field drains or rubble drains may overcome this problem.

6.2.4 Existing drains

As buildings must not be constructed over an existing drain (including a field drain) that is to remain active, before any building work is carried out, a survey should be carried out to establish the geography and topography of the building site to ascertain whether there are any existing field drains.

If an active subsoil drain is cut during excavation and if it passes under the building it should be either: • relaid in pipes with sealed joints and have access points outside the building; or • rerouted around the building; or • rerun to another outfall (see Figure 6.2).	BR-AD-C3.3

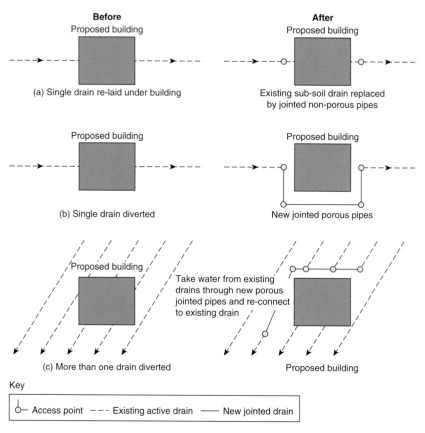

Before
Proposed building

(a) Single drain re-laid under building

After
Proposed building

Existing sub-soil drain replaced
by jointed non-porous pipes

Proposed building

(b) Single drain diverted

Proposed building

New jointed porous pipes

Proposed building

Take water from existing
drains through new porous
jointed pipes and re-connect
to existing drain

(c) More than one drain diverted

Proposed building

Key

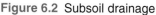

⊙— Access point   – – – Existing active drain   —— New jointed drain

Figure 6.2 Subsoil drainage

Where a building site requires that an existing drain (including a field drain) must remain active and be rerouted, or retained, particular methods of construction and protection should be carefully considered.	SBR-D-3.5.1, SBR-ND-3.5.1

Buildings must not be constructed over an existing drain (including a field drain) that is to remain active.

Where it is proposed to construct a building over the line of an existing sewer, the sewer should be rerouted around the building.	SBR-D-3.5.0, SBR-ND-3.5.0

Permission will be required from the water authority for any work that is to be carried out to a public sewer or where it is necessary to build over a public sewer.

A survey should be carried out to establish the geography and topography of the building site and ascertain whether there are any existing field drains.	SBR-D-3.5.1, SBR-ND-3.5.1
Where a building site requires that an existing drain (including a field drain) must remain active and be rerouted or retained particular methods of construction and protection should be carefully considered.	SBR-D-3.5.1, SBR-ND-3.5.1
Where a building is erected over a private drain, including a field drain that is to remain active, the drain should be rerouted if reasonably practicable or reconstructed in a manner appropriate to the conditions of the site.	SBR-D-3.5.2, SBR-ND-3.5.2

In non-domestic buildings, it would be unreasonable for drains to be rerouted around a limited-life building, but care should be taken that no undue loading is transmitted to the drain that might cause damage.

Where a drain or sewer passes another drain or sewer or goes through, under or close to structures (including a manhole or inspection chamber) a detail should be devised to allow sufficient flexibility to avoid damage of the pipe due to movement.	SBR-D-3.5.4, SBR-ND-3.5.4
Disused sewers or drains (which provide ideal nesting sites for rats) should be disconnected from the drainage system as near as possible to the point of connection.	SBR-D-3.5.5, SBR-ND-3.5.5
Sewers and drains less than 1.5 m from the surface and in open ground should be, as far as reasonably practicable, removed. Other pipes should be capped at both ends and at any point of connection, to ensure rats cannot gain entry.	SBR-D-3.5.5, SBR-ND-3.5.5

6.2.5 Water fittings

Any water fitting laid below ground level *shall* have a depth of cover sufficient to prevent water freezing in the fitting.	SI 1148-7(4)
Any concealed water fitting or mechanical backflow prevention device, not being a terminal fitting, *will* be made of gunmetal, or another material resistant to dezincification.	SI 1148-7(3)

Note: 'Concealed water fitting' means a water fitting which:

- is installed below ground;
- passes through or under any wall, footing or foundation;
- is enclosed in any chase or duct; or
- is in any other position which is inaccessible or renders access difficult.

7

Drainage

One of the main areas for consideration when constructing a building, altering a building, putting up an extension or in any other way installing a water fitting is to consider the disposal of rainwater, foul water and brown water.

7.1 What are the rules about drainage?

The Building Acts require that all drains are either connected with a sewer (unless the sewer is more than 120 feet (36.5 m) away or the person carrying out the building work is not entitled to have access to the intervening land) or are able to discharge into a private wastewater treatment plant or septic tank, cesspool, settlement tank or other tank specifically designed for the reception and/or disposal of foul matter from buildings.

The local authorities view this requirement very seriously and will need to be satisfied that:

- satisfactory provision has been made for drainage;
- all cesspools, private sewers, septic tanks, drains, soil pipes, rainwater pipes, spouts, sinks or other appliances are adequate for the building in question;
- all private sewers that connect directly or indirectly to the public sewer are incapable of admitting subsoil water;
- the condition of a cesspool is not detrimental to health, and does not present a nuisance;
- cesspools, private sewers and drains previously used, but now no longer in service, do not prejudice health or become a nuisance.

These requirements can become quite a problem if they are not recognized in the early planning stages and so it is always best to seek the advice of the local authority. In certain circumstances, the local authority may even help to pay for the cost of connecting you up to the nearest sewer.

The local authority has the power to make the owner renew, repair or cleanse existing cesspools, sewers and drains, etc.

7.1.1 Can two buildings share the same drainage?

Usually the local authority will require every building to be drained separately into an existing sewer but in some circumstances they may decide that it would be more cost effective if the buildings were drained in combination. On occasions, they may even recommend that a private sewer is constructed.

7.1.2 What about ventilation of soil pipes?

A major requirement of the Building Regulations is that all soil pipes from water closets *shall* be properly ventilated and that:

- no existing or proposed pipe that is designed to carry rainwater from a roof is used to convey soil and drainage from a sanitary convenience;
- no existing pipe that is designed to carry surface water from a premises is used as a ventilating shaft for a drain or a sewer conveying foul water.

7.1.3 What happens if I need to disconnect an existing drain?

If, in the course of your building work, you need to:

- reconstruct, renew or repair an existing drain that is joined up with a sewer or another drain;
- alter the position of an existing drain that is joined up with a sewer or another drain;
- seal off an existing drain that is joined up with a sewer or another drain;

then, provided that you give 48 hours' notice to the local authority, the person undertaking the reconstruction may break open any street for this purpose.

You do not need to comply with this requirement if you are demolishing an existing building.

7.1.4 Can I repair an existing water closet or drain?

Repairs may be carried out to water closets, drains and soil pipes, but if that repair or construction work is prejudicial to health and/or could be construed as being a public nuisance, then the person who completed the installation or repair is liable, on conviction, to a heavy fine.

Note: In some areas, a 'water closet' can also be taken to mean a urinal.

7.1.5 Can I repair an existing drain?

You are required by the Building Act 1984 to ensure that all courts, yards and passageways that provide access to a house, industrial or commercial building (not maintained at public expense) are capable of allowing satisfactory drainage of its surface or subsoil to a proper outfall. Only in *extreme* emergencies, however, are you allowed to repair, reconstruct or alter the course of an underground drain that joins up with a sewer, cesspool or other drainage method (e.g. septic tank) without first having permission from the local authority.

If you have to carry out repairs, etc., in an emergency, then make sure that you do not cover over the drain or sewer without notifying the local authority so that they can (if they wish) inspect your work.

7.1.6 Are there separate systems for drainage?

Separate systems of drains and sewers *shall* be provided for foul water and rainwater where:

- the rainwater is not contaminated; and
- the drainage is to be connected either directly or indirectly to the public sewer system, which has separate systems for foul water and surface water.

7.2 Requirements

Every building must be designed and constructed in such a way that there will not be a threat to the building or the health of occupants or people in or around the building:

- as a result of flooding and the accumulation of groundwater;
- as a result of moisture penetration from the ground.

All plans for building work need to show that drainage of refuse water (e.g. from sinks) and rainwater (from roofs) have been adequately catered for. Failure to do so will mean that these plans will be rejected by the local authority.

All plans for buildings must include at least one (or more) water or earth closets unless the local authority are satisfied that one is not required (for example in a large garage separated from the house).

If you propose using an earth closet, the local authority cannot reject the plans unless they consider that there is insufficient water supply to that earth closet.

Buildings must not be constructed over an existing drain (including a field drain) that is to remain active.

Any system for discharging water to a sewer shall be separate from that provided for the conveyance of foul water from the building.

This requirement only applies to a system that is provided in connection with the erection or extension of a building where it is reasonably practicable for the system to discharge directly or indirectly to a sewer for the separate conveyance of surface water.

Every drain or sewer should be protected (e.g. by providing barriers) from damage by construction traffic and heavy machinery.

Heavy materials should not be stored over drains or sewers.

Manholes should not be located within a dwelling.

After laying (including any necessary concrete or other haunching or surrounding and backfiring) gravity drains and private sewers should be tested for water tightness.

Material alterations to existing drains and sewers are subject to (and covered by) the Building Regulations.

Repairs, reconstruction and alterations to existing drains and sewers should be carried out to the same standards as new drains and sewers.

Drains should provide a degree of fire protection.

Note: this is normally provided by ensuring that:

- all openings in fire-separating elements shall be suitably protected in order to maintain the integrity of the continuity of the fire separation;
- any hidden voids in the construction shall be sealed and subdivided to inhibit the unseen spread of fire and products of combustion, in order to reduce the risk of structural failure, and the spread of fire.

7.2.1 Surface water drainage

Every building, and hard surface within the curtilage of a building, must be designed and constructed with a surface water drainage system that will:

- ensure the disposal of surface water without threatening the building and the health and safety of the people in and around the building; and
- have facilities for the separation and removal of silt, grit and pollutants.

7.2.2 Rainwater drainage

Adequate provision shall be made for rainwater to be carried from the roof of the building.

Paved areas around the building shall be so constructed as to be adequately drained.

Rainwater from a system provided to meet the above requirements shall discharge to one of the following, listed in order of priority:

- an adequate soakaway or some other adequate infiltration system; or, where that is not reasonably practicable,
- watercourse; or, where that is not reasonably practicable,
- a sewer.

Rainwater drainage systems shall:

- minimize the risk of blockage or leakage;
- be accessible for clearing blockages;

- ensure that rainwater soaking into the ground is distributed sufficiently so that it does not damage foundations of the proposed building or any adjacent structure;
- ensure that rainwater from roofs and paved areas is carried away from the surface either by a drainage system or by other means;
- ensure that the rainwater drainage system carries the flow of rainwater from the roof to an outfall (e.g. a soakaway, a watercourse, surface water or a combined sewer).

 Note: Moisture from precipitation penetrating to the inner face of the building shall not pose a threat to either the structure of the building or the health of the occupants.

7.2.3 Wastewater drainage

Every wastewater drainage system serving a building must be designed and constructed in such a way as to ensure the removal of wastewater from the building without threatening the health and safety of the people in and around the building, and;

- that facilities for the separation and removal of oil, fat, grease and volatile substances from the system are provided;
- that discharge is to a public sewer or public wastewater treatment plant, where it is reasonably practicable to do so; and
- where discharge to a public sewer or public wastewater treatment plant is not reasonably practicable that discharge is to a private wastewater treatment plant or septic tank.

7.2.4 Foul water drainage

'Foul water' means wastewater which removes the waste from the toilet, bidet, bath, basins, sinks, washing machines, dishwashers and showers. Surface water drainage deals with rainfall as it collects around your property. In older houses the surface water is often fed into the foul water system.

 Foul water is not allowed to be fed into a surface water system and all underground pipes must be brown in colour to distinguish them from any other underground service pipes.

General requirements

Every building must be designed and constructed in such a way that the foul water drainage system shall:

- convey the flow-off foul water to a foul water outfall (i.e. sewer, cesspool, septic tank or settlement (i.e. holding) tank);
- minimize the risk of blockage or leakage;

- prevent foul air from the drainage system from entering the building under working conditions,
- be ventilated;
- be accessible for clearing blockages.
- not increase the vulnerability of the building to flooding.

An adequate system of drainage shall be provided to carry foul water from appliances within the building to one of the following, listed in order of priority:

- a public sewer;
- a private sewer communicating with a public sewer;
- a septic tank which has an appropriate form of secondary treatment;
- a wastewater treatment system; or
- a cesspool.

7.2.5 Wastewater treatment systems and cesspools

Wastewater treatment systems shall:

- have sufficient capacity to enable breakdown and settlement of solid matter in the wastewater from the buildings;
- be sited and constructed so as to prevent overloading of the receiving water.

Cesspools shall have sufficient capacity to store the foul water from the building until they are emptied.

Wastewater treatment systems and cesspools shall be sited and constructed so as not to:

- be prejudicial to health or a nuisance;
- adversely affect water sources or resources;
- pollute controlled waters;
- be in an area where there is a risk of flooding.

Septic tanks and wastewater treatment systems and cesspools shall be constructed and sited so that they:

- will not be prejudicial to the health of any person;
- will not contaminate any watercourse, underground water or water supply;
- will have adequate ventilation;
- prevent leakage of the contents and ingress of subsoil water;
- have regard to water table levels at any time of the year and rising groundwater levels;
- have adequate means of access for emptying and maintenance; and
- where relevant, will function to a sufficient standard for the protection of health in the event of a power failure.

Drainage fields shall be sited and constructed so as to:

- avoid overloading of the soakage capacity; and provide adequately for the availability of an aerated layer in the soil at all times.

7.2.6 Existing sewers

Building or extension work involving underpinning shall:

- be constructed or carried out in a manner which will not overload or otherwise cause damage to the drain, sewer or disposal main either during or after the construction;
- not obstruct reasonable access to any manhole or inspection chamber on the drain, sewer or disposal main;
- in the event of the drain, sewer or disposal main requiring replacement, not unduly obstruct work to replace the drain, sewer or disposal main, on its present alignment;
- reduce the risk of damage to the building as a result of failure of the drain, sewer or disposal main;
- provide adequately for the availability of an aerated layer in the soil at all times.

7.2.7 Septic tank and holding tanks

Any septic tank, holding tank which is part of a wastewater treatment system or cesspool shall be:

- of adequate capacity;
- so constructed that it is impermeable to liquids; and
- adequately ventilated.

Where a foul water drainage system from a building discharges to a septic tank, wastewater treatment system or cesspool, a durable notice shall be affixed in a suitable place in the building containing information on any continuing maintenance required to avoid risks to health.

7.2.8 Private wastewater treatment plant or septic tank

Every private wastewater treatment plant or septic tank serving a building must be designed and constructed so that:

- it will ensure the safe temporary storage and treatment of wastewater prior to discharge;
- the disposal of the wastewater to ground is safe and is not a threat to the health of the people in or around the building.

7.2.9 Solid waste storage

Every building must be designed and constructed in such a way that accommodation for solid waste storage does not contaminate any water supply, groundwater or surface water.

7.3 Meeting the requirements

7.3.1 Surface water drainage

Every building, and hard surface within the curtilage of a building, must be designed and constructed with a surface water drainage system that will:

- ensure the disposal of surface water without threatening the building and the health and safety of the people in and around the building; and
- have facilities for the separation and removal of silt, grit and pollutants.

Climate change is expected to result in more rain in the future and it is essential that this is taken into account in today's buildings and that all surface water from buildings is:

- removed quickly and safely without damage to the building or danger to people around the building, and does not pose a risk to the environment by flooding or pollution;
- cleared quickly from all access routes to buildings, particularly with elderly and disabled people in mind.

Some drainage authorities have sewers that carry both foul water and rainwater (i.e. combined systems) in the same pipe. Where they do, they can allow rainwater to discharge into the system if the sewer has enough capacity to take the added flow.

Some private sewers (drains serving more than one property) also carry both foul water and rainwater. If a sewer (or private sewer) operated as a combined system does not have enough capacity, the rainwater should be run in a separate system with its own outfall.

Discharge to a watercourse may require consent from the Environment Agency, which may limit the rate of discharge. Where other forms of outlet are not practicable, discharge should be made to a sewer. For design purposes a rainfall interval of 0.014 litres/second/m^2 can be assumed as normal.

Under section 85 (offences concerning the polluting of controlled waters) of the Water Resources Act 1991 it is an offence to discharge any noxious or polluting material into a watercourse, coastal water or underground water. Most surface water sewers discharge to watercourses.

Under section 111 (restrictions on use of public sewers) of the Water Industry Act 1991 it is an offence to discharge petrol into any drain or sewer connected to a public sewer.

Every building, and hard surface within the curtilage | SBR-D-3.6,
of a building, must be designed and constructed with a | SBR-ND-3.6
surface water drainage system that will:

- ensure the disposal of surface water without
  threatening the building and the health and safety
  of the people in and around the building; and
- have facilities for the separation and removal of
  silt, grit and pollutants.

Surface water drainage should discharge to a | BR-AP-H3.2
soakaway or other infiltration system where
practicable.

Surface water drainage connected to combined sewers | BR-AP-H3.7,
should: | BR-AP-H3.14

- have traps on all inlets:
- be at least 75 mm diameter.

Where any materials that could cause pollution are | BR-AP-H3.21
stored or used, separate drainage systems should be
provided.

Note: On car parks, petrol filling stations or other areas where there is likely to be leakage or spillage of oil, drainage systems should be provided with oil interceptors. Separators should be leak tight and comply with the requirements of the Environmental Agency and prEN858.

Infiltration devices (including soakaways, swales, | BR-AP-
infiltration basins and filter drains) should not be | H3.23–26
built:

- within 5 m of a building or road or in areas of
  unstable land;
- in ground where the water table reaches the
  bottom of the device at any time of the year;
- close to any drainage fields, drainage mounds
  or other soakaways;
- where the presence of any contamination
  in the run-off could result in pollution of a
  groundwater source or resource.

To avoid ponding of water on paved surfaces, particularly in winter where ice can form, paved surfaces that are accessible to pedestrians should be drained quickly and efficiently.

A paved surface, such as a car park, of less than 200 m² is unlikely to contribute to flooding problems and may be designed to have free-draining run-off.

Every building should be provided with a drainage system that is capable of removing surface water from paved surfaces (particularly an access route used by disabled people) without endangering the building or the health and safety of people in and around the dwelling.	SBR-D-3.6.2, SBR-ND-3.6.2
Paved surfaces should be so laid so that the rainwater run-off is not close to the building.	SBR-D-3.6.2, SBR-ND-3.6.2

Sustainable urban drainage systems

Sustainable urban drainage systems (SUDS) are built to manage surface water run-off and are used in conjunction with good management of the land to prevent pollution.

A SUDS technique for surface water drainage should be provided in accordance with the guidance contained in 'Sustainable Urban Drainage Systems: design manual for Scotland and Northern Ireland'.	SBR-D-6.4, SBR-ND-6.4

The maintenance of a SUDS within the curtilage of a building is the responsibility of the building owner

Surface water discharge

Surface water that is discharged from a building and/or a hard surface within the curtilage of a building should be carried to a disposal point that will not endanger the building, the environment or the health and safety of people in and around the building.

Surface water discharge should be to:

• a SUDS;	SBR-D-3.6.3,
• a soakaway;	SBR-ND-3.6.3
• a public sewer;	
• an outfall to a watercourse, such as a river, stream or loch or coastal waters; or to	
• a storage container with an overflow discharging into any of the four options above.	

Surface water run-off from small paved areas should:

- be free draining to a pervious area, such as grassland, provided the soakage capacity of the ground is not overloaded.   SBR-D-3.6.6, SBR-ND-3.6.6
- not discharge adjacent to a building where it could damage the foundations.

Drainage system outside a building

A drainage system outside a dwelling should be constructed and installed in accordance with the recommendations in BS EN 12056-1:2000, BS EN 752-3:1997 (amendment 2), BS EN 752-4:1998 and BS EN 1610:1998.   SBR-D-3.7.3, SBR-ND-3.7.3

Drains passing through structures

Where a drain or sewer passes through, under or close to another drain or sewer and/or structure (including a manhole or an inspection chamber) the design should allow sufficient flexibility so as to avoid damage of the pipe due to movement.   SBR-D-3.5.4, SBR-ND-3.5.4

Rerouting of drains

If there is no alternative and a building has to be constructed over the line of an existing sewer, the sewer should be rerouted around the building.

If a building is erected over a private drain, including a field drain that is to remain active, the drain should be rerouted or reconstructed, in a manner appropriate to the conditions of the site.   SBR-D-3.5.2, SBR-ND-3.5.2

In non-domestic buildings, it is normally considered unreasonable for drains to be rerouted around a limited-life building, nevertheless care should be taken that no undue loading is transmitted to the drain that may cause damage.

Reconstruction of drains

During construction, it should be ensured that the assumptions made in the design are safeguarded or adapted to changed conditions.

Every drain or sewer should be protected (e.g. by providing barriers) from damage by construction traffic and heavy machinery.	SBR-D-3.5.3, SBR-ND-3.5.3
Heavy materials should not be stored over drains or sewers.	SBR-D-3.5.3, SBR-ND-3.5.3

Sealing disused drains

Disused sewers or drains (which provide ideal nesting sites for rats) should be disconnected from the drainage system as near as possible to the point of their connection.	SBR-D-3.5.5, SBR-ND-3.5.5
Sewers and drains less than 1.5 m from the surface and in open ground should be, as far as reasonably practicable, removed. Other pipes should be capped at both ends and at any point of connection, to ensure rats cannot gain entry.	SBR-D-3.5.5, SBR-ND-3.5.5

Soakaways

Soakaways have been the traditional method of disposal of surface water from buildings and paved areas where no mains drainage exists. They should not, however, endanger the stability of the building.

Every part of a soakaway should be located at least 5 m from a building and from a boundary.	SBR-D-3.6.3, SBR-ND-3.6.3
A soakaway serving a single dwelling, small building or an extension should be tested in accordance with the percolation test method.	SBR-D-3.6.5, SBR-ND-3.6.5
There should be individual soakaways for each building.	SBR-D-3.6.5, SBR-ND-3.6.5
Soakaways less than 100 m^2 in total area shall consist of square or circular pits filled with rubble or lined with dry jointed masonry or perforated ring units. Soakaways for larger areas shall be lined pits or trench type soakaways.	BR-AP-H3.26
The storage volume should be calculated so that, over the duration of a storm, it is sufficient to contain the difference between the inflow volume and the outflow volume.	BR-AP-H3.29

Soakaways serving larger areas should be designed in accordance with BS EN 752.   BR-AP-H3.30

Provisions for washing down

Where communal solid waste storage is located within a building, such as where a refuse chute is utilized, the storage area should have provision for washing down and draining the floor into a wastewater drainage system.

Every building must be designed and constructed in such a way that accommodation for solid waste storage is provided which does not contaminate any water supply, groundwater or surface water.	SBR-D-3.25
Gullies should incorporate a trap that maintains a seal even during periods of disuse.	SBR-D-3.25.4
Walls and floors should be of an impervious surface that can be washed down easily and hygienically.	SBR-D-3.25.4
The enclosures should be permanently ventilated at the top and bottom of the wall.	SBR-D-3.25.4

Discharges into a drainage system from car wash and vehicle repair facilities

Where a discharge into a traditional drainage system contains silt or grit, for example from a hard standing with car wash facilities: SBR-D-3.6.9, SBR-ND-3.6.9

- there should be facilities for the separation of such substances;
- removable grit interceptors should be incorporated into the surface water gully pots to trap the silt or grit.

Where a discharge into a drainage system contains oil, grease or volatile substances (e.g. from a vehicle repair garage) there should be facilities for the separation and removal of such substances.   SBR-ND-3.6.9

The use of emulsifiers to break up any oil or grease in the drain is not recommended as they can cause problems further down the system.

Dungsteads and farm effluent tanks

Every dungstead or farm effluent tank, including a slurry or silage effluent tank, should be constructed in such a manner so as to prevent the escape of effluent through the structure that could cause ground contamination or environmental pollution.	SBR-D-3.26.1, SBR-ND-3.26.1
The construction should also prevent seepage and overflow that might endanger any water supply or watercourse.	SBR-D-3.26.1, SBR-ND-3.26.1
Covers or fencing should be in accordance with the relevant recommendations of Section 8 of BS 5502: Part 50:1993.	SBR-D-3.26.3, SBR-ND-3.26.3

7.3.2 Rainwater drainage

Precipitation

Rain penetration occurs most often through walls exposed to the prevailing wet winds and unless there are adequate damp proof courses and flashings, etc., materials in parapets and chimneys can collect rainwater and deliver it to other parts of the dwelling below roof level.

Every building must be designed and constructed in such a way that there will not be a threat to the building, the occupants and/or the health of people in or around the building as a result of moisture from precipitation penetrating to the inner face of the building.	SBR-D-3.10, SBR-ND-3.10

This requirement does not apply to a building where penetration of moisture from the outside will result in effects no more harmful than those likely to arise from use of the building.

A floor, wall, roof or other building element exposed to precipitation, or wind-driven moisture, should prevent penetration of moisture to the inner surface of any part of a building so as to protect the occupants and to ensure that the building is not damaged.	SBR-D-3.10.1, SBR-ND-3.10.1
Any water fitting conveying rainwater, recycled water or any fluid other than water supplied by a water undertaker, shall be clearly identified so as to be easily distinguished from any supply pipe or distributing pipe.	SI 1148-14(1)(a) 14(1)(b)

No supply pipe, distributing pipe or pump   SI 1148-14(2)
delivery pipe that draws water from a sup-
ply pipe or distributing pipe shall convey (or
be connected so that it can convey) any fluid
unless a device for preventing backflow is
installed.

Surface rainwater drainage

Every building should be provided with a drainage system to remove rainwater from the roof (or other areas where rainwater might accumulate) without causing damage to the structure or endangering the health and safety of people in and around the building.

Methods other than gutters and rainwater pipes are normally used.

Gutters and rainwater pipes may be omitted from a   SBR-D-3.6.1,
roof at any height provided it has an area of not more   SBR-ND-3.6.1
than 8 m^2 and no other area drains onto it.
If an eaves drop system that allows rainwater to drop   SBR-D-3.6.1,
freely to the ground is used, it should be designed so   SBR-ND-3.6.1
that it:

- protects the fabric of the dwelling from ingress
  of water caused by water splashing on the wall;
- prevents water from entering doorways and
  windows;
- protects persons from falling water when
  around the dwelling;
- protects persons and the building fabric from
  rainwater splashing on the ground or forming
  ice on access routes;
- protects building foundations from concentrated
  discharges from gutters.

Drainage of paved areas

Surface gradients should direct water draining from   BR-AP-H3 (2.2)
a paved area away from buildings.
Gradients on impervious surfaces should be designed   BR-AP-H3 (2.3)
to permit the water to drain quickly from the surface.

Note: A gradient of at least 1 in 60 is recommended.

Paths, driveways and other narrow areas of paving should be free draining to a pervious area such as grassland, provided that:	BR-AP-H3 (2.6)
• the water is not discharged adjacent to buildings where it could damage foundations; and • the soakage capacity of the ground is not overloaded.	
Where water is to be drained onto the adjacent ground the edge of the paving should be finished above or flush with the surrounding ground to allow the water to run off.	BR-AP-H3 (2.7)
Where the surrounding ground is not sufficiently permeable to accept the flow, filter drains may be provided.	BR-AP-H3 (2.8 and 3.33)
Pervious paving should not be used where excessive amounts of sediment are likely to enter the pavement and block the pores.	BR-AP-H3 (2.11)
Pervious paving should not be used in oil storage areas, or where run-off may be contaminated with pollutants.	BR-AP-H3 (2.12)
Gullies should be provided at low points where water would otherwise pond.	BR-AP-H3 (2.15)
Gully gratings should be set approximately 5 mm below the level of the surrounding paved area in order to allow for settlement.	BR-AP-H3 (2.16)
Provision should be made to prevent silt and grit entering the system, either by provision of gully pots of suitable size, or catchpits.	BR-AP-H3 (2.17)

Rainwater drainage

The capacity of the drainage system should be large enough to carry the expected flow at any point in the system.	BR-AP-H3 (0.3)
Rainwater or surface water should not be discharged to a cesspool or septic tank.	BR-AP-H3 (0.6)

Gutters and rainwater pipes

Although this part of the Building Regulations only applies to draining the rainfall from areas of $8\,m^2$ or more (unless they receive a flow from a rainwater pipe or from paved and/or other hard surfaces), each case should be considered separately and a decision made. This particularly applies to small roofs and balconies. Table 7.1 shows the largest effective area that should be drained into the gutter sizes most often used.

Table 7.1 **Gutter and outlet sizes**

Max. effective roof area (m²)	Gutter size (mm diameter)	Outlet size (mm diameter)	Flow capacity (l/s)
6.0	–	–	–
18.0	75	50	0.38
37.0	100	63	0.78
53.0	115	63	1.11
65.0	125	75	1.37
103.0	150	89	2.16

For eaves gutters the design rainfall intensity should be $0.021\,litres/second/m^2$. In some cases, eaves drop systems may be used.

Gutters should be laid with any fall towards the nearest outlet.

Gutters should be laid so that any overflow in excess of the design capacity (e.g. above normal rainfall) will be discharged clear of the building.	BR-AP-H3 (1.7)
Rainwater pipes should discharge into a drain or gully (but may discharge to another gutter or onto another surface if it is drained).	BR-AP-H3 (1.8)
Any rainwater pipe which discharges into a combined system should do so through a trap.	BR-AP-H3 (1.8)
The size of a rainwater pipe should be at least the size of the outlet from the gutter.	BR-AP-H3 (1.10)
A down pipe which serves more than one gutter should have an area at least as large as the combined areas of the outlets.	BR-AP-H3 (1.10)
On flat roofs, valley gutters and parapet gutters additional outlets may be necessary.	BR-AP-H3 (1.7)
Where a rainwater pipe discharges onto a lower roof or paved area, a pipe shoe should be fitted to divert water away from the building.	BR-AP-H3 (1.9)
Gutters and rainwater pipes should be firmly supported without restricting thermal movement.	

The materials used should be of adequate strength   BR-AP-H3 (1.16)
and durability, and

- all gutter joints should remain watertight
  under working conditions;
- pipework in siphonic roof drainage systems
  should be able to resist to negative pressures
  in accordance with the design;
- gutters and rainwater pipes should be firmly
  supported;
- different metals should be separated by
  non-metallic material to prevent electrolytic
  corrosion.

Rainwater harvesting

Rainwater harvesting systems allow surface water run-off from dwellings or hard standing areas to be collected, processed, stored and distributed, thereby reducing the demand for potable water, the load on drainage systems and surface water run-off that can lead to incidents of flooding.

Rain, as it falls on buildings, is soft, clear and largely free of contaminants. During collection and storage, however, there is potential for contamination. For this reason it is recommended that recycled surface water is used only for flushing water closets, car washing and garden taps.

Water storage tanks should be constructed of   SBR-D-3.6.7,
materials such as glass reinforced plastic (GRP),   SBR-ND-3.6
high-density polyethylene, steel or concrete which
is sealed and protected against the corrosive effects
of the stored water and to prevent the ingress of
groundwater if located underground.

Prior to the storage of water in a tank the rainwater SBR-D-3.6.7,
should be filtered to remove leaves and other organic SBR-ND-3.6
matter, dust and/or grit.

Disinfection may be required if the catchment area   SBR-D-3.6.7,
is liable to be contaminated with animal faeces,   SBR-ND-3.6
extensive bird droppings, oils or soil.

Water for use in the building should be extracted from SBR-D-3.6.7,
just below the water surface in the tank to provide   SBR-ND-3.6
optimum water quality.

All pipework carrying rainwater for use in the building should be identified as such in accordance with the Water Regulations Advisory Scheme (WRAS) guidance notes and great care should be taken to avoid cross-connecting reclaimed water and mains water.	SBR-D-3.6.7, SBR-ND-3.6
Tanks should be accessible to allow for internal cleaning and the maintenance of inlet valves, sensors, filters or submersible pumps.	SBR-D-3.6.7, SBR-ND-3.6
An overflow should discharge to a soakaway or to mains drainage where it is not reasonably practicable to discharge to a soakaway.	SBR-D-3.6.7, SBR-ND-3.6
Backflow prevention devices should be incorporated to prevent contaminated water from entering the system.	SBR-D-3.6.7, SBR-ND-3.6
A surface water drainage system should be tested to ensure the system is laid and is functioning correctly.	SBR-D-3.6.10, SBR-ND-3.6.10

7.3.3 Wastewater drainage

Where discharge to a public sewer or public wastewater treatment plant is not reasonably practicable discharge should be to a private wastewater treatment plant or septic tank.

A careful check should be made before breaking into an existing drain to ensure it is the correct one (e.g. whether wastewater to surface water or vice versa) and a further test carried out after connection, such as a dye test, to confirm correct connection.

Every wastewater drainage system serving a building must be designed and constructed so that all wastewater is removed from the building without threatening the health and safety of the people in and around the building, and that:	SBR-D-3.7, SBR-ND-3.7

• facilities for the separation and removal of oil, fat, grease and volatile substances from the system are provided;
• discharge is to a public sewer or public wastewater treatment plant, where it is reasonably practicable to do so.

> Water supply systems shall be capable of being   SI 1148-11
> drained down and be fitted with an adequate
> number of servicing valves and drain taps so as to
> minimize the discharge of water when water fittings
> are maintained or replaced. A sufficient number of
> stopvalves shall be installed for isolating parts of the
> pipework.

Note: Some sewers, called combined sewers, carry wastewater and surface water in the same pipe. In some cases it may be appropriate to install a drainage system within the curtilage of a building as a separate system even when the final connection is to a combined sewer.

Manholes

Health and safety legislation requires that manual entry to a drain or sewer system is only undertaken where no alternative exists and that remotely operated equipment is the normal method of access.

Manholes should not be located within a dwelling.	SBR-D-3.5.3, SBR-ND-3.5.3
Drainage gratings and manhole covers should not create a trip or entrapment hazard.	SBR-D-4.1.4, SBR-ND-4.1.4

Where a private drain discharges into a public sewer (normally at the curtilage of a building) access should be provided for maintenance and this is usually via a disconnecting inspecting chamber (Figure 7.1) immediately inside the curtilage of the building. It is preferable that this chamber is provided for individual houses but where this is not practicable, a shared disconnecting chamber may be provided.

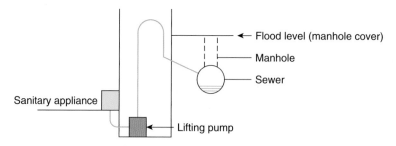

Figure 7.1 Disconnecting chamber

A disconnecting chamber (or manhole where the depth is more than 1.2 m) should be provided in accordance with the requirements of local water authority.

SBR-D-3.7.4, SBR-ND-3.7.4

The disconnecting chamber, or manhole, for a block of individually owned flats or maisonettes should be located as close to the building as is reasonably practicable as the drain will become a public sewer once it passes outside the footprint of the building.

SBR-D-3.7.4

Ideally, a separate drainage system carrying wastewater and surface water should be constructed within the curtilage of a building.

SBR-D-3.7.5

 Note: Some combined sewers carry wastewater and surface water in the same pipe. These systems are not recommended nowadays as they are more likely to surcharge during heavy rains, but in older properties they still sometimes exist.

Any water fitting conveying:

SI 1148-14(1)(a)

- rainwater, recycled water or any fluid other than water supplied by a water undertaker; or
- any fluid that is not wholesome water;

14(1)(b)

shall be clearly identified so as to be easily distinguished from any supply pipe or distributing pipe.

No supply pipe, distributing pipe or pump delivery pipe drawing water from a supply pipe or distributing pipe shall convey, or be connected so that it can convey, any fluid falling within sub-paragraph 14(1) unless a device for preventing backflow is installed with sub-paragraph 15.

Domestic buildings

A wastewater drainage system should discharge to a public sewer or public wastewater treatment plant, where it is reasonably practicable to do so.

SBR-D-3.7.9

Non-domestic buildings

> Discharge of grey water may be via a water closet   SBR-ND-3.7.11
> into a public sewer or public wastewater treatment
> plant, where it is reasonably practicable to do so.

In Scotland, a wastewater drainage system should discharge to a public sewer or public wastewater treatment plant provided under the Sewerage (Scotland) Act 1968, where it is reasonably practicable to do so.

> A wastewater drainage system serving a building   SBR-D-3.7.7,
> should be ventilated to limit the pressure fluctuations   SBR-ND-3.7.7
> within the system and minimize the possibility of foul
> air entering the building.

Wastewater treatment systems and cesspools

A notice giving information as to the nature and frequency of maintenance required for the cesspool or wastewater treatment system to continue to function satisfactorily should be displayed within each of the buildings.

The use of non-mains foul drainage, such as wastewater treatment systems, septic tanks or cesspools, should only be considered where connection to mains drainage is not practicable.

Any discharge from a wastewater treatment system is likely to require consent from the Environment Agency. For the detailed design and installation of small sewage treatment works, specialist knowledge is advisable. Guidance is also given in BS 6297:1983, *Code of Practice for Design and Installation of Small Sewage Treatment Works and Cesspools*.

Septic tanks

Septic tanks with some form of secondary treatment (such as from a drainage field/mound or constructed wetland such as a reed bed) will normally be the most economic means of treating wastewater from small developments (e.g. one to three dwellings). They provide suitable conditions for the settlement, storage and partial decomposition of solids which need to be removed at regular intervals.

> Septic tanks should be sited at least 7 m from any   BR-AP-
> habitable parts of buildings, and preferably down a slope.   H1(1.16)

Septic tanks should only be used in conjunction with a form of secondary treatment (e.g. a drainage field, drainage mound or constructed wetland).	BR-AP-H1(1.15)
Septic tanks should be sited within 30 m of a vehicle access to enable the tank to be emptied and cleaned without hazard to the building occupants and without the contents being taken through a dwelling or place of work.	BR-AP-H1(1.17 and 1.64)
Septic tanks and settlement tanks should have a capacity below the level of the inlet of at least 2700 litres (2.7 m³) for up to four users. This size should be increased by 180 litres for each additional user.	BR-AP-H1(1.18)
Septic tanks may be constructed in brickwork or concrete (roofed with heavy concrete slabs) and/or factory-manufactured septic tanks (made out of glass reinforced plastics, polyethylene or steel) can be used.	BR-AP-H1(1.19–20 and 1.65–66
The brickwork should consist of engineering bricks at least 220 mm thick. The mortar should be a mix of 1:3 cement sand ratio and in situ concrete should be at least 150 mm thick of C/25/P mix (see BS 5328).	BR-AP-H1(1.120 and 1.66)
Septic tanks should: • be ventilated; • incorporate at least two chambers or compartments operating in series; • be provided with access for emptying and cleaning; • be inspected monthly to check they are working correctly; • be emptied at least once a year.	BR-AP-H1(1.21–1.24), BR-AP-H1(A.11 and A.13
A notice should be fixed within the building describing the necessary maintenance.	BR-AP-H1(1.25)

Cesspools

A cesspool is a watertight tank, installed underground, for the storage of sewage. No treatment is involved.

A filling rate of 150 litres per person per day is assumed and if the cesspool does not fill within the estimated period, the tank should be inspected for leakage.

Cesspools should be:	BR-AP-H2 1.58, 1.60, 1.63 and A.21R
• sited at least 7 m from any habitable parts of buildings and preferably downslope;	
• provided with access for emptying and cleaning;	
• emptied on a monthly basis by a licensed contractor;	
• ventilated.	
The inlet of a cesspool should be provided with access for inspection.	BR-AP-H2 (1.67)
Cesspools and settlement tanks (if they are to be desludged using a tanker) should be sited within 30 m of a vehicle access.	BR-AP-H2 (1.64)
Cesspools and settlement tanks should prevent leakage of the contents and ingress of subsoil water.	BR-AP-H2 (1.63)
Cesspools should have a capacity below the level of the inlet of at least 18,000 litres (18 m³) for two users, increased by 6800 litres (6.8 m³) for each additional user.	BR-AP-H2 (1.61)
Cesspools, septic tanks and settlement tanks may be constructed in brickwork, concrete or glass reinforced concrete.	BR-AP-H2 (1.65–66)

Note: Factory-made cesspools and septic tanks are available in GRP, polyethylene or steel.

The brickwork should consist of engineering bricks at least 220 mm thick. The mortar should be a mix of 1:3 cement sand ratio and in situ concrete should be at least 150 mm thick of C/25/P mix (see BS 5328). Cesspools should be covered (with heavy concrete slabs) and ventilated.	BR-AP-H2 (1.66)
Cesspools should:	BR-AP-H2 (1.62) BR-AP-H2 (A.20)
• have no openings except for the inlet, access for emptying and ventilation;	
• be inspected fortnightly for overflow and emptied as required.	

Packaged treatment works

This term is applied to a range of systems designed to treat a given hydraulic and organic load using prefabricated components which can be installed with minimal site work. They are capable of treating effluent more efficiently than septic tank systems and this normally allows the product to be directly discharged to a watercourse.

The discharge from the wastewater treatment plant should be sited at least 10 m away from watercourses and any other buildings.	BR-AP-H2 (1.54)
Regular maintenance and inspection should be carried out in accordance with the manufacturer's instructions.	BR-AP-H2 (A.17)

Drainage fields and mounds

Drainage fields (or mounds) serving a wastewater treatment plant or septic tank should be located:	BR-AP-H2 (1.27)

- at least 10 m from any watercourse or permeable drain;
- at least 50 m from the point of abstraction of any groundwater supply;
- at least 15 m from any building;
- sufficiently far from any other drainage fields, drainage mounds or soakaways so that the overall soakage capacity of the ground is not exceeded.

No water supply pipes or underground services other than those required by the disposal system itself should be located within the disposal area.	BR-AP-H2 (1.29)
No access roads, driveways or paved areas should be located within the disposal area.	BR-AP-H2 (1.30)
The ground water table should not rise to within 1 m of the invert level of the proposed effluent distribution pipes.	BR-AP-H2 (1.33)
An inspection chamber should be installed between the septic tank and the drainage field.	BR-AP-H2 (1.43)
Constructed wetlands should not be located in the shade of trees or buildings.	BR-AP-H2 (1.47)
The drainage field/mound should be checked on a monthly basis to ensure that it is not waterlogged and that the effluent is not backing up towards the septic tank.	BR-AP-H2 (A.15)

Under section 50 (overflowing and leaking cesspools) of the Public Health Act 1936 action could be taken against a builder who had caused the problem, and not just against the owner.

Under section 59 (drainage of building) of the Building Act 1984, local authorities can require either the owner or the occupier to remove (or otherwise make innocuous) any disused cesspool, septic tank or settlement tank.

Wastewater drainage systems: conversions and extensions

A careful check should be made before breaking into an existing drain to ensure it is the correct one (e.g. whether wastewater to surface water or vice versa) and a further test carried out after connection, such as a dye test, to confirm correct connection.	SBR-D-3.7.6

7.3.4 Foul water drainage

General requirements

The capacity of the system should be large enough to carry the expected flow at any point (BS 5572, BS 8301).	BR-AP-H1-0.1
All pipes, fittings and joints should be capable of withstanding an air test of positive pressure of at least 38 mm water gauge for at least 3 minutes.	BR-AP-H1-1.38
Every trap should maintain a water seal of at least 25 mm.	BR-AP-H1-1.38

Foul drainage

Some public sewers may carry foul water and rainwater in the same pipe. If the drainage system is also to carry rainwater to such a sewer these combined systems should not be capable of discharging into a cesspool or septic tank. Foul drainage should be connected to either:	BR-AP-H1-2.1
• a public foul or combined sewer (wherever this is reasonably practicable;	BR-AP-H1-2.3
• an existing private sewer that connects with a public sewer; or	BR-AP-H1-2.6

• a wastewater treatment system or cesspool should be provided.	BR-AP-H1-2.7
Combined and rainwater sewers shall be designed to surcharge (i.e. the water level in the manhole rises above the top of the pipe) in heavy rainfall.	BR-AP-H1-2.8
Basements containing sanitary appliances, where the risk of flooding due to sewer surcharge of the sewer is possible should use either an anti-flooding valve (if the risk is low) or be pumped. For other low-lying sites (i.e. not basements) where the risk is considered low, a gully (at least 75 mm below the floor level) can be dug outside the building.	BR-AP-H1- 2.9, 2.10 and 2.36–2.39
Anti-flooding valves should preferably be a double valve type that complies with prEN 13564.	BR-AP-H1- 2.11
The layout of the drainage system should be kept simple. Pipes should (wherever possible) be laid in straight lines. Changes of direction and gradient should be minimized.	BR-AP-H1- 2.13
Access points should be provided only if blockages could not be cleared without them.	BR-AP-H1- 2.13
Connections should be made using prefabricated components.	BR-AP-H1- 2.15
Connection of drains to other drains or private or public sewers and of private sewers to public sewers should be made obliquely, or in the direction of flow.	BR-AP-H1- 2.14
The system should be ventilated by a flow of air.	BR-AP-H1- 1.27–1.29 and 2.18
Ventilating pipes should not finish near openings in buildings.	BR-AP-H1- 1.31 and 2.18
Pipes should be laid to even gradients and any change of gradient should be combined with an access point.	BR-AP-H1- 2.19 and 2.49
Pipes should also be laid in straight lines where practicable.	BR-AP-H1- 2.20 and 2.49
Where foul water drainage from a building is to be pumped, the effluent receiving chamber should be sized to contain 24-hour inflow to allow for disruption in service.	BR-AP-H1 (2.39)
The minimum daily discharge of foul drainage should be taken as 150 litres per head per day for domestic use.	BR-AP-H1 (2.39)

Discharge pipes

Branch pipes should discharge into either another branch pipe or a discharge stack (unless the appliances discharge into a gully on the ground floor or at basement level).	BR-AP-H1-15.5
If the appliances are on the ground floor, the pipe(s) may discharge to a stub stack, discharge stack, directly to a drain or (if the pipe carries only wastewater) to a gully.	BR-AP-H1-1.5–1.17 and 1.30
A branch pipe from a ground floor closet should only discharge directly to a drain if the depth from the floor to the drain is 1.3 m or less (see Figure 7.2).	BR-AP-H1-1.9

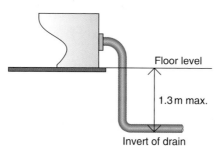

Figure level

1.3 m max.

Invert of drain

Figure 7.2 Direct connection of ground floor WC to drain

A branch pipe serving any ground floor appliance may discharge direct to a drain or into its own stack.	BR-AP-H1-A5
A branch pipe should not discharge into a stack in a way which could cause cross-flow into any other branch pipe (see Figure 7.3).	BR-AP-H1-1.10

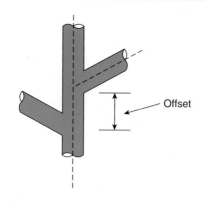

Offset

Figure 7.3 Branch connections

A branch discharge pipe should not discharge into a stack lower than 450 mm above the invert of the tail of the bend at the foot of the stack in single dwellings up to three storeys (see Figure 7.4).

BR-AP-H1-1.8 and 1.21, A3, A4

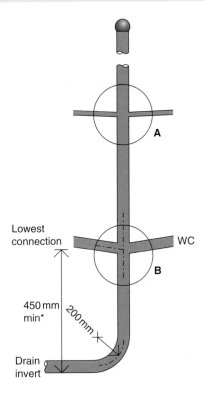

Figure 7.4 Branch discharge stack

Branch pipes may discharge into a stub stack.

BR-AP-H1-12 and 1.30

A branch pipe discharging to a gully should terminate between the grating or sealing plate and the top of the water seal.

BR-AP-H1-1.13

Bends in branch pipes should be avoided if possible. Junctions on branch pipes should be made with a sweep of 25 mm radius or at 45°.

BR-AP-H1-1.16

BR-AP-H1-1.17

Rodding points should be provided to give access to any lengths of discharge pipes which cannot be reached by removing traps or appliances with integral traps.

BR-AP-H1-1.6 and 1.25

A branch pipe discharging to a gully should terminate   BR-AP-H1-1.13
between the grating or sealing plate and the top of the
water seal.

Pipes serving a single appliance should have at least the same diameter as the
appliance trap (Table 7.2).

Table 7.2 **Minimum trap sizes and seal depths**

Appliance	Diameter of trap (mm)	Depth of seal (mm of water or equivalent)
Washbasin	32	75
Bidet		
Bath	40	50
Shower		
Food waste disposal unit	40	75
Urinal bowl		
Sink	75	50
Washing machine		
Dishwashing machine		
WC pan (outlet < 80 mm)		
WC pan (outlet > 80 mm)	100	50

A separate ventilating stack is only likely to be preferred where there are a
number of sanitary appliances and their distance to a discharge stack is large.

Building over existing sewers

Where it is proposed to construct a building over or   BR-AP-H4 (0.3)
near a drain or sewer shown on any map of sewers,
the developer should consult the owner of the drain
or sewer.

A building constructed over or within 3 m of any: BR-AP-H4 (1.2)

- rising main;
- drain or sewer constructed from brick or
  masonry;
- drain or sewer in poor condition;

shall not be constructed in such a position unless
special measures are taken.

Buildings or extensions should not be constructed   BR-AP-H4 (1.3)
over a manhole or inspection chamber or other
access fitting on any sewer (serving more than one
property).

A satisfactory diversionary route should be available so that the drain or sewer could be reconstructed without affecting the building.	BR-AP-H4 (1.4)
The length of drain or sewer under a building should not exceed 6 m except with the permission of the owners of the drain or sewer.	BR-AP-H4 (1.5)
Buildings or extensions should not be constructed over or within 3 m of any drain or sewer more than 3 m deep, or greater than 225 mm in diameter except with the permission of the owners of the drain or sewer.	BR-AP-H4 (1.60)
Where a drain or sewer runs under a building at least 100 mm of granular or other suitable flexible filling should be provided round the pipe.	BR-AP-H4 (1.9)
Where a drain or sewer running below a building is less than 2 m deep, the foundation should be extended locally so that the drain or sewer passes through the wall.	BR-AP-H4 (1.10)
Where the drain or sewer is more than 2 m deep to invert and passes beneath the foundations, the foundations should be designed with a lintel spanning over the line of the drain or sewer. The span of the lintel should extend at least 1.5 m either side of the pipe and should be designed so that no load is transmitted onto the drain or sewer.	BR-AP-H4 (1.12)
A drain trench should not be excavated lower than the foundations of any building nearby.	BR-AP-H4 (1.13)

7.3.5 Wastewater treatment systems and cesspools

A wastewater treatment system is an effective, economical way of treating wastewater from buildings. It consists of two main components, a watertight underground tank into which raw sewage is fed and a system designed to discharge the wastewater safely to the environment without pollution.

This discharge is normally an infiltration field through which wastewater is released to the ground, but when ground conditions are not suitable, a discharge to a watercourse or coastal waters may be permitted.

Wastewater treatment systems shall:

- have sufficient capacity to enable breakdown and settlement of solid matter in the wastewater from the buildings;
- be sited and constructed so as to prevent overloading of the receiving water.

Wastewater treatment systems and cesspools shall be sited and constructed so as not to:

- be prejudicial to health or a nuisance;
- adversely affect water sources or resources;
- pollute controlled waters;
- be in an area where there is a risk of flooding.

Septic tanks and wastewater treatment systems and cesspools shall be constructed and sited so that they:

- will not contaminate any watercourse, underground water or water supply;
- have adequate means of access for emptying and maintenance; and
- where relevant, will function to a sufficient standard for the protection of health in the event of a power failure;
- will have adequate ventilation;
- prevent leakage of the contents and ingress of subsoil water;
- have regard to water table levels at any time of the year and rising groundwater levels.

Drainage fields shall be sited and constructed so as to:

- avoid overloading of the soakage capacity; and provide adequately for the availability of an aerated layer in the soil at all times.

Existing sewers

Building or extension work involving underpinning shall:

- be constructed or carried out in a manner which will not overload or otherwise cause damage to the drain, sewer or disposal main either during or after the construction;
- not obstruct reasonable access to any manhole or inspection chamber on the drain, sewer or disposal main;
- in the event of the drain, sewer or disposal main requiring replacement, not unduly obstruct work to replace the drain, sewer or disposal main, on its present alignment;
- reduce the risk of damage to the building as a result of failure of the drain, sewer or disposal main.
- provide adequately for the availability of an aerated layer in the soil at all times.

Septic tank and holding tanks

Any septic tank or holding tank which is part of a wastewater treatment system or cesspool shall be:

- of adequate capacity;
- so constructed that it is impermeable to liquids; and
- adequately ventilated.

Where a foul water drainage system from a building discharges to a septic tank, wastewater treatment system or cesspool, a durable notice shall be affixed in a suitable place in the building containing information on any continuing maintenance required to avoid risks to health.

7.3.6 Private wastewater treatment plant or septic tank

Where it is not reasonably practicable to connect to a public sewer or a public wastewater treatment plant then discharge should be to a private wastewater treatment system (plant or septic tank).

Although the domestic use of detergents and disinfectants may not necessarily be detrimental, excessive use may have a harmful effect on the performance of the sewage treatment works.

All private wastewater treatment plants or septic tanks serving a building must be designed and constructed so that they will ensure the safe temporary storage and treatment of wastewater prior to discharge.	SBR-D-3.8, SBR-ND-3.8
A private wastewater treatment plant and septic tank should be designed, constructed and installed in accordance with:	SBR-D-3.8.1, SBR-ND-3.8.1

- the recommendations of BS EN 12566-1:2000, for a prefabricated septic tank; or
- the recommendations of BS 6297:1983; or
- the conditions of certification by a notified body.

The settlement tank of a private wastewater plant and a septic tank should have a securely sealed, solid cover that is capable of being opened by one person using standard operating keys.	SBR-D-3.8.2, SBR-ND-3.8.2

Private wastewater infiltration systems

Wastewater from treatment systems can discharge either to land via an infiltration system (Figure 7.5) or to watercourses, lakes, lochs or coastal waters.

Every private wastewater treatment system serving a building must be designed and constructed in such a way that the disposal of the wastewater to ground is safe and is not a threat to the health of the people in and/or around the building.	SBR-D-3.9, SBR-ND-3.9

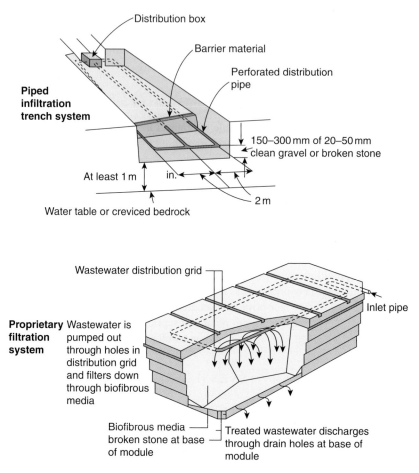

Figure 7.5 Types of filtration system

Location of a private wastewater treatment plant

Research has shown that there are no specific health issues that dictate a safe location of a treatment plant or septic tank relative to a building. However, damage to the foundations of a building has been shown to occur where leakage from the tank has occurred. Thus, in the unlikely event of there being leakage, it is sensible to ensure that any water-bearing strata direct any effluent away from the dwelling. To prevent any such damage, therefore:

Every part of a private wastewater plant and septic tank should be located: • at least 5 m from a building; • at least 5 m from a boundary.	SBR-D-3.8.4, SBR-ND-3.8.4

Assessing the suitability of the ground

An infiltration system serving a private wastewater treatment plant, septic tank or for grey water discharge should be constructed in ground suitable for the treatment and dispersion of the wastewater being discharged.

A ground assessment and soil percolation test should   SBR-D-3.9.1,
be carried out to determine the suitability of the   SBR-ND-3.9.1
ground using the three-step procedure described in
Figure 7.6 and Table 7.3.

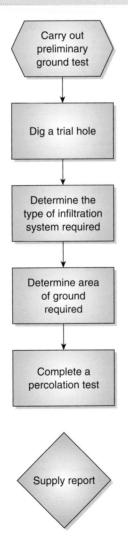

Figure 7.6 Ground assessment and soil percolation test

Table 7.3 **Ground assessment and soil percolation test**

1	Carry out a preliminary ground assessment	• Consult Water Authority (SEPA and verifier in Scotland) and the Environmental Health Officer as required; • consult Water Authority's (SEPA in Scotland) latest groundwater protection policy; • identify underlying geology and aquifers; • identify whether the ground is liable to flooding; • nature of the subsoil and groundwater vulnerability; • implication of plot size; • proximity of underground services; • ground topography and local drainage patterns; • identify whether water is abstracted for drinking, used in food processing or farm dairies; • implication for, and of, trees and other vegetation; • location of surface waters and terrestrial ecosystems
2	Dig a trial hole to determine the position of the water table and soil conditions	The trial hole should be: • a minimum of 2 m deep; or • a minimum of 1.5 m below the invert of the proposed distribution pipes; and • left covered for a period of 48 hours before measuring any water table level
3	Determine the type of infiltration system and the area of ground required	Carry out a percolation test

Design of infiltration fields

An infiltration system serving a private wastewater treatment plant or septic tank should be designed and constructed to suit the conditions as determined by the ground into which the treated wastewater is discharged.

Location of infiltration fields

An infiltration system serving a private wastewater treatment plant or septic tank should be located so as to minimize the risk of pollution and/or damage to buildings.

An infiltration field should be located:	SBR-D-3.9.4
• at least 50 m from any spring, well or borehole used for resupplying drinking water;	SBR-D-3.9.4 and SBR-D-3.9.5

- at least 10 m horizontally from any watercourse   SBR-ND-3.9.5
  (including any inland or coastal waters),
  permeable drain, road or railway;
- at least 5 m from a building;
- at least 5 m from the boundary.

Access for desludging: private wastewater treatment plant

Where provided, a private wastewater treatment plant and septic tank should
have reasonable access for desludging.

Wastewater treatment plants should be inspected monthly to check they are working correctly.	SBR-D-3.8.6, SBR-ND-3.8.6
The effluent in the outlet from the tank should be free flowing.	SBR-D-3.8.6, SBR-ND-3.8.6
The frequency of desludging will depend upon the capacity of the tank and the amount of waste draining to it from the building.	SBR-D-3.8.6, SBR-ND-3.8.6

If you require advice on desludging frequencies, speak to the tank manufacturer or the desludging contractor.

The desludging tanker should be provided with access   SBR-D-3.8.6,
to a working area that: SBR-ND-3.8.6

- will provide a clear route for the suction hose from
  the tanker to the tank;
- is not more than 25 m from the tank if it is not more
  than 4 m higher than the invert level of the tank; and
- is sufficient to support a vehicle axle load of
  14 tonnes.

Inspection and sampling

A private wastewater plant and septic tank should be provided with a chamber for the inspection and sampling of the wastewater discharged from the tank.	SBR-D-3.8.3, SBR-D-3.8.3
The owner should carry out inspections at regular intervals and a chamber should be provided in accordance with Figure 7.7.	SBR-D-3.8.3, SBR-D-3.8.3

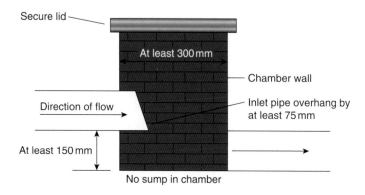

Secure lid

At least 300 mm

Chamber wall

Direction of flow

Inlet pipe overhang by
at least 75 mm

At least 150 mm

No sump in chamber

Figure 7.7 Section through inspection chamber

In Scotland, where mains drainage is frequently unavailable and discharges of sewage effluent is to ground via an infiltration system or to a watercourse, loch or coastal waters, then Scottish Environmental Protection Agency (SEPA) permission is required.

Labelling: private wastewater treatment plant or septic tank

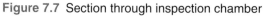

Every building with a drainage system discharging to a private wastewater treatment plant or septic tank should be provided with a label to alert the occupiers to such an arrangement.	SBR-D-3.8.7, SBR-ND-3.8.7
The label should describe the recommended maintenance necessary for the system and should include the wording shown in Figure 7.8.	SBR-D-3.8.7, SBR-ND-3.8.7

'The drainage system from this property discharges to a wastewater treatment plant (or septic tank, as appropriate).

The owner is legally responsible for routine maintenance and to ensure that the system complies with any discharge consent issued by SEPA and that it does not present a health hazard or a nuisance'.

Figure 7.8 Labelling of private wastewater treatment plant

The label should be located adjacent to the gas or electricity consumer unit or the water stopcock.	SBR-D-3.8.7, SBR-ND-3.8.7

7.3.7 Drainage from sanitary facilities

The drainage from sanitary facilities in dwellings is of prime concern and every building must be designed and constructed in such a way that sanitary facilities are provided for all occupants of, and visitors to, the building in a form that allows convenience of use and that there is no threat to the health and safety of occupants or visitors.

Note: Although not recommending that sanitary facilities on the principal living level of a dwelling be designed to an optimum standard for wheelchair users, it should be possible for most people, no matter their disability, to use these facilities unassisted and in privacy.

Every dwelling shall have sanitary facilities comprising at least one water closet (WC), or waterless closet, together with one wash hand basin per WC, or waterless closet, one bath or shower and one sink.	SBR-D-3.12.1

Sanitary appliances below flood level

The basements of approximately 500 buildings in Scotland are flooded each year when the sewers surcharge (i.e. when the effluent runs back up the pipes because they are too small to take the required flow).

Wastewater from sanitary appliances and floor gullies below flood level should be drained by wastewater lifting plants (constructed in accordance with the requirements of BS EN 12056-4:2000) or, where there is unlikely to be a risk to persons such as in a car park, via an anti-flooding device.	SBR-D-3.7.2, SBR-ND-3.7.2
Wastewater from sanitary appliances above flood level should not be drained through anti-flooding devices and only in special case, e.g. refurbishment, by a wastewater lifting plant.	SBR-D-3.7.2, SBR-ND-3.7.2

7.3.8 Grey water

Water reuse is becoming an accepted method of reducing demand on mains water and the use of grey water may be appropriate in some buildings for flushing of water closets. However, because grey water recycling systems require constant observation and maintenance they should only be used in buildings where a robust maintenance contract exists.

In Scotland, the approval of Scottish Water is required before any such scheme is installed.

Where a grey water system is to be used, it should be designed, installed and commissioned by a person competent and knowledgeable in the nature of the system and the regulatory requirements.	SBR-ND-3.7.9
A risk assessment on the health and safety implications should be carried out for those who will be employed to install and maintain the system.	SBR-ND-3.7.9
A comprehensive installation guide, users' guide and an operation and maintenance manual should be handed to the occupier at the commissioning stage.	SBR-ND-3.7.9
All pipework carrying grey water for reuse should be clearly marked with the word 'GREY WATER'.	BR-AP-H1(A11)

Grey water and rainwater tanks

Grey water and rainwater tanks should:	BR-AP-H2 (1.70)
• prevent leakage of the contents and ingress of subsoil water;	
• be ventilated;	
• have an anti-backflow device;	
• be provided with access for emptying and cleaning.	

Grey water disposal

The disposal of grey water (from baths, showers, washbasins, sinks and washing machines) may be accomplished by an infiltration field.	SBR-D-3.9.3, SBR-ND-3.9.3

The area of an infiltration field can be calculated using the following formula:

$$A = P \times Vp \times 0.2$$

where A is the area of the subsurface drainage trench (m^2), P is the number of people served, and Vp is the percolation value obtained (mm/s).

7.3.9 Solid waste storage

Every building must be designed and constructed in such a way that accommodation for solid waste storage does not contaminate any water supply, groundwater or surface water.

Although the requirements of the Building Regulations do not cover the recycling of household and other waste, they do cover the general requirements for solid waste storage.

Where a discharge into a drainage system contains oil, fat, grease or volatile substances (e.g. from a commercial kitchen) there should be facilities for the separation and removal of such substances.	SBR-ND-3.7.8
The use of emulsifiers to break up any oil or grease in the drain is not recommended as they can cause problems further down the system.	SBR-ND-3.7.8
A wastewater drainage system should be tested to ensure the system is laid and is functioning correctly.	SBR-D-3.7.8, SBR-ND-3.7.10
Communal storage areas should have provision for washing down and draining the floor into a system suitable for receiving a polluted effluent.	BR-AP-H4 (1.14)
Gullies should incorporate a trap that maintains a seal even during prolonged periods of disuse.	BR-AP-H4 (1.14)

7.3.10 Fuel storage

Oil is a common and highly visible form of water pollution because of the way it spreads; even a small quantity can cause a lot of harm to the aquatic environment. Oil can pollute rivers, lakes, lochs, groundwater and coastal waters, killing wildlife and removing vital oxygen from the water.

The UK government is required by the European Water Framework Directive (2006/118/EC) to prevent List I substances from entering groundwater and to prevent groundwater pollution by List II substances (EC Dangerous Substances Directive (76/464/EEC).

Every building must be designed and constructed in such a way that an oil storage installation, incorporating oil storage tanks used solely to serve a fixed combustion appliance installation providing space heating or cooking facilities in a building, will:	SBR-D-3.24, SBR-ND-3.24

- reduce the risk of oil escaping from the installation;
- contain any oil spillage likely to contaminate any water supply, groundwater, watercourse, drain or sewer; and permit any spill to be disposed of safely;
- minimize the number of journeys by delivery vehicles.

LPG storage: fixed tanks

A liquefied petroleum gas (LPG) storage tank, together with any associated pipework, should be designed, constructed and installed in accordance with the requirements set out in the Liquefied Petroleum Gas Association (LPGA) Code of Practice 1: *Bulk LPG Storage at Fixed Installations*.

Note: LPG storage tanks in excess of 4 tonnes (9000 litres) capacity are uncommon in domestic applications.

Externally located, above-ground, oil tanks with a capacity of not more than 2500 litres serving a building should be provided with a catchpit or be integrally bunded if the tank is:	SBR-D-3.24.3, SBR-ND-3.24.30

- located within 10 m of the water environment (i.e. rivers, lochs, coastal waters);
- located where spillage could run into an open drain or to a loose-fitting manhole cover;
- within 50 m of a borehole or spring;
- over ground where conditions are such that oil spillage could run off into a watercourse;
- located in a position where the vent pipe outlet is not visible from the fill point.

In a domestic building, secondary containment should be provided where a tank is within a building or wholly below ground.	SBR-D-3.24.3

Ground features such as open drains, manholes, gullies and cellar hatches should be sealed or trapped to prevent the passage of LPG vapour.

7.3.11 Rodent control

If the site has been previously developed, the local authority should be consulted to determine whether any special measures are necessary for the control of rodents. Special measures which may be taken include the following:

Sealed drainage – should have access covers to the pipework in the inspection chamber instead of an open channel.	BR-AP-H1 (2.22a)
Intercepting traps – should be of the locking type that can be easily removed from the chamber surface and securely replaced.	BR-AP-H1 (2.22b)
Rodent barriers – including enlarged sections on discharge stacks to prevent rats climbing, flexible downward facing fins in the discharge stack, or one-way valves in underground drainage.	BR-AP-H1 (2.22c)
Metal cages on ventilator stack terminals – to discourage rats from leaving the drainage system.	BR-AP-H1 (2.22d and 1.31)
Covers and gratings to gullies – used to discourage rats from leaving the system.	BR-AP-H1 (2.22e)
During construction, drains and sewers that are left open should be covered when work is not in progress to prevent entry by rats.	BR-AP-H1 (2.56)
Disused drains or sewers less than 1.5 m deep that are in open ground should as far as is practicable be removed. Other pipes should be sealed at both ends (and at any point of connection) and grout filled to ensure that rats cannot gain access.	BR-AP-H1 (B18)

7.3.12 Drain pipes

Pipe gradients and sizes

Drains should have enough capacity to carry the anticipated maximum flow (see Table 7.4).	BR-AP-H1 (2.29)

Table 7.4 **Flow rates from dwellings**

No. of dwellings	Flow rate (l/s)
1	2.5
5	3.5
10	4.1
15	4.6
20	5.1
26	5.4
30	5.8

Sewers (i.e. a drain serving more than one property) should have a minimum diameter of 100 mm when serving fewer than 10 dwellings or a diameter of 150 mm if more than 10.	BR-AP-H1 (2.30)
Drains carrying foul water should have an internal diameter of at least 75 mm.	BR-AP-H1 (2.33)
Drains carrying effluent from a WC or trade effluent should have an internal diameter of at least 100 mm.	BR-AP-H1 (2.33)

To minimize the effects of any differential settlement, pipes should have flexible joints.	BR-AP-H1 (2.40)
All joints should remain watertight under working and test conditions.	BR-AP-H1 (2.40)
Nothing in the pipes, joints or fittings should project into the pipe line or cause an obstruction.	BR-AP-H1 (2.40)
Different metals should be separated by non-metallic materials to prevent electrolytic corrosion.	H1 (2.40)

Materials for pipes and jointing

A drain may run under a building if at least 100 mm of granular or other flexible filling is provided round the pipe.	BR-AP-H1 (2.23)
Where pipes are built into a structure (e.g. inspection chamber, manhole, footing, ground beam or wall) suitable measures (such as using rocker joints or a lintel) should be taken to prevent damage or misalignment (see Figures 7.9 and 7.10).	BR-AP-H1 (2.24)

Protection from settlement

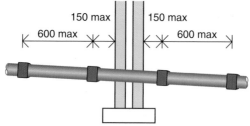

Figure 7.9 Pipe embedded in the wall. Short length of pipe bedded in a wall with joints 150 mm of either wallface. Additional rocker pipes (maximum length 600 mm) with flexible joints are then added

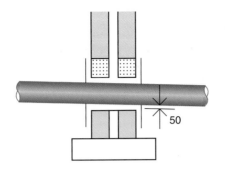

Figure 7.10 Pipe shielded by a lintel. Both sides are masked with rigid sheet material (to prevent entry of fill or vermin) and the void is filled with a compressible sealant to prevent entry of gas

The depth of cover will usually depend on the levels of the connections to the system, the gradients at which the pipes should be laid and the ground levels. All drain trenches should not be excavated lower than the foundations of any building nearby (see Figure 7.11).	BR-AP-H1 (2.27) and H1 (2.41–2.45) BR-AP-H1 (2.25)

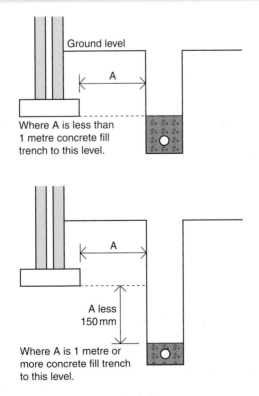

Figure 7.11 Pipe running near a building

Trenches may be backfilled with concrete to protect nearby foundations. In these cases a movement joint (as shown in Figure 7.12) formed with a compressible board should be provided at each socket or sleeve joint.

BR-AP-H1
(2.45)

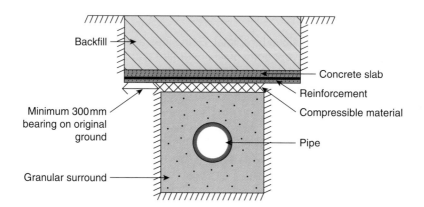

Figure 7.12 Joints for concrete-encased pipes

Bedding and backfill

The choice of bedding and backfill depends on the depth at which the pipes are to be laid and the size and strength of the pipes.

Special precautions should be taken to take account of the effects of settlement where pipes run under or near buildings.

The depth of the pipe cover will usually depend on the levels of the connections to the system and the gradients at which the pipes should be laid and the ground levels.

Pipes need to be protected from damage, particularly pipes which could be damaged by the weight of backfilling.

BR-AP-H1
(2.41)

Rigid pipes should be laid in a trench as shown in Figure 7.13.

BR-AP-H1
(2.42)

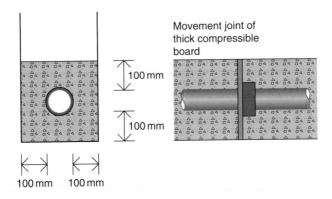

Figure 7.13 Bedding for rigid pipes

Flexible pipes shall be supported to limit deformation under load.	BR-AP-H1 (2.44)
Flexible pipes with very little cover shall be protected from damage by a reinforced cover slab with flexible filler and at least 75 mm of granular material between the top of the pipe and the underside of the flexible filler below the slabs (see Figure 7.14).	BR-AP-H1 (2.44)

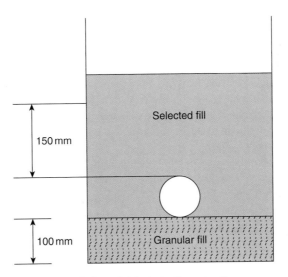

Figure 7.14 Protection of pipes laid at shallow depths

Access points

Access should be provided to long runs.	BR-AP-H1 (2.50)
Sufficient and suitable access points should be provided for clearing blockages from drain runs that cannot be reached by any other means.	BR-AP-H1 (2.46)
Access points should be provided:	BR-AP-H1 (2.49)

- on or near the head of each drain run;
- at a bend;
- at a change of gradient or pipe size;
- at a junction.

Access points should be either:	BR-AP-H1 (2.48)

- rodding eyes – capped extensions of the pipes;
- access fittings – small chambers on (or an extension of) the pipes but not with an open channel;
- inspection chambers – chambers with working space at ground level;
- manholes – deep chambers with working space at drain level.

Access points should be constructed so as to resist the ingress of groundwater or rainwater.	BR-AP-H1 (2.52)
Inspection chambers and manholes should have removable non-ventilating covers of durable material (such as cast iron, cast or pressed steel, precast concrete or unplasticized polyvinyl chloride (uPVC)).	BR-AP-H1 (2.54)
Access points to sewers (serving more than one property) should be located in places where they are accessible and apparent for use in an emergency (e.g. highways, public open space, unfenced front gardens, unfenced driveways).	BR-AP-H1 (2.51)
Inspection chambers and manholes in buildings should have mechanically fixed airtight covers unless the drain itself has watertight access covers.	BR-AP-H1 (2.54)
Manholes deeper than 1 m should have metal step irons or fixed ladders.	BR-AP-H1 (2.54)

8

Ventilation

There shall be adequate means of ventilation provided for people in
the building.

(BR-AP-F)

Ventilation is defined in the Building Regulations as 'the supply and removal
of air (by natural and/or mechanical means) to and from a space or spaces in
a building'.

In addition to replacing 'stale' indoor air with 'fresh' outside air, from the
point of view of water, the aim of ventilation is also to:

- limit the accumulation of moisture from a building which could, otherwise, become a health hazard to people living and/or working within that
  building; and
- control excess humidity.

In general terms, both of these aims can be met if the ventilation system:

- disperses residual water vapour;
- extracts water vapour from wet areas where it is produced in significant
  quantities (e.g. kitchens, utility rooms and bathrooms);
- rapidly dilutes water vapour produced in habitable rooms, occupiable
  rooms and sanitary accommodation;
- provides protection against rain penetration.

One method of achieving good indoor air quality is to reduce the amount of
water vapour and/or air pollutants released into the indoor air, particularly
those from construction and consumer products. Ventilation air intakes that
are located on a less polluted side of the building may be used for fresh air.

Note: Further information about control of emissions from construction products is available in BRE Digest 464.

8.1 Ventilation systems

Extract ventilation concerns the removal of air directly from a space or spaces to outside and this may be achieved by natural means, such as passive stack ventilation (PSV) (Figure 8.1), or by mechanical means, such as an extract fan or a central system.

The following systems may be used in dwellings without basements.

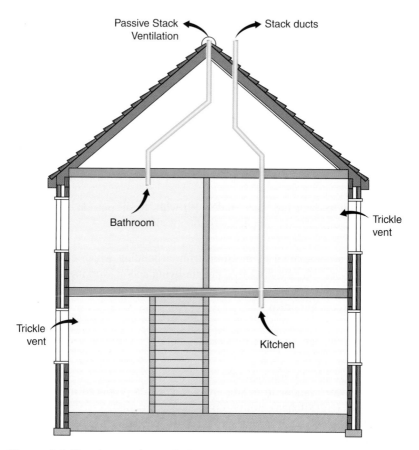

Figure 8.1 Passive stack ventilation

8.1.1 Passive stack ventilation

This is a ventilation device that uses ducts between terminals mounted in the ceiling of a room(s) and terminals on the roof to extract air to the outside, by a combination of the natural movement of air through the building and the difference in temperature and pressure between the inside of the building and the effects of wind passing over the roof of the building outside (Figures 8.2 and 8.3). As a result, the moist air from within the building is drawn up through the ducts and replaced by fresh air that flows through trickle vents mounted in the wall or as part of the window frame in habitable rooms.

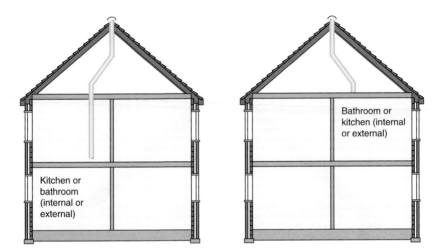

Figure 8.2 Preferred passive stack ventilation system layouts

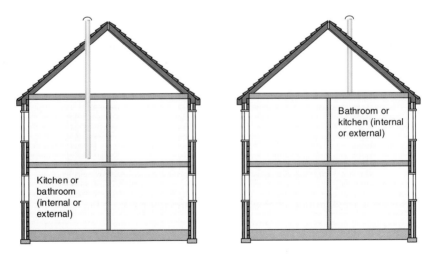

Figure 8.3 Alternative passive stack ventilation system layouts

The so-called 'stack effect' relies on the pressure differential between the inside and the outside of a building caused by differences in the density of the air caused by an indoor/outdoor temperature difference.

8.1.2 Background ventilators

Background ventilation (Figure 8.4) is a combination of uncontrolled air infiltration and ventilation through vents that are designed to dilute the unavoidable contaminant emissions from people and materials (e.g. radon) and to control the levels of internal moisture due to occupant activities. In turn, this will reduce the amount of moisture in the air that could otherwise lead to mould growth and other pollutants. The need for background ventilators will depend on the air permeability or air tightness of a building.

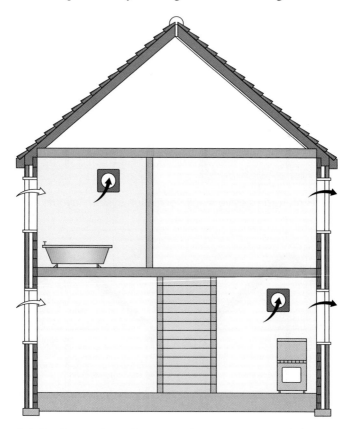

Figure 8.4 Background ventilators and intermittent extract fans

Note: Air permeability is defined as the average volume of air (in cubic metres/hour) that passes through unit area of the building envelope (in square metres) when subject to an internal to external pressure difference of 50 Pa.

8.1.3 Continuous mechanical extract

This system may consist of either a central extract system or individual room fans, or a combination of both (Figure 8.5).

Continuous mechanical extract

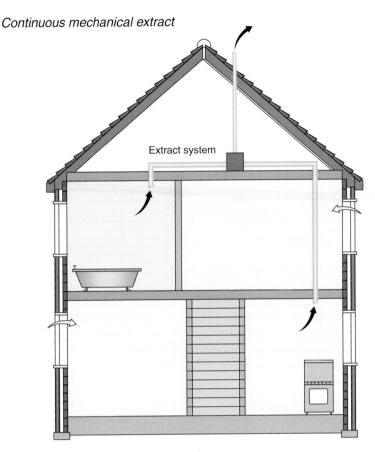

Figure 8.5 Continuous mechanical extract

The three fan types most commonly used in domestic applications are:

- **axial fans** – the most common form of fan, which can be mounted on a wall, in a window (i.e. through a suitable glazing hole) or in the ceiling (such as in a bathroom);
- **centrifugal fans** – which (because they develop greater pressure) permit longer lengths of ducting to be used and so can be used for most wall and/ or window applications in high-rise (i.e. above three storeys) buildings or in exposed locations so as to overcome wind pressure;
- **in-line fans** – which need the shortest possible duct length to the discharge terminal, unless they are in-line mixed flow fans which have the characteristics of both axial and centrifugal fans and can, therefore, be used with longer lengths of ducting.

 Note: In-line and in-line mixed fans can be used for bathrooms (100 mm diameter), utility rooms (125 mm diameter) and kitchens (150 mm diameter).

8.1.4 Trickle ventilators

Manually controlled trickle ventilators are widely used for background ventilation. They can be located as shown in Figure 8.6.

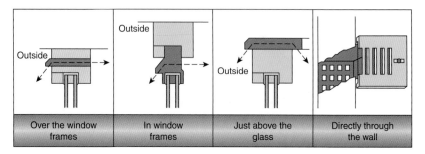

| Over the window frames | In window frames | Just above the glass | Directly through the wall |

Figure 8.6 Trickle ventilation system

To avoid cold draughts, trickle ventilators are normally positioned 1.7 m above floor level and usually include a simple control (such as a flap) to allow users to shut off the ventilation according to personal choice or external weather conditions. Nowadays, pressure-controlled trickle ventilators that reduce the air flow according to the pressure difference across the ventilator are available to reduce draught risks during windy weather. Trickle ventilators are normally left open in the occupied room of a dwelling.

8.1.5 Purge ventilation

Purge ventilation is a manually controlled type of ventilation that is used in rooms and spaces to dilute water vapour rapidly. It can be achieved by natural means (e.g. an openable window or an external door) or by mechanical means (e.g. a fan) (Figure 8.7).

Figure 8.7 Ventilating fan and trunk hose

For sanitary accommodation only, purge ventilation may be used provided that security is not an issue.

8.1.6 Night purge ventilation

During the day the heat gains caused through solar energy and the occupants of the building are stored in the thermal mass of any exposed concrete, brickwork and/or stone used for the walls, floor or ceiling. At night this heat is released through vents or open windows.

Note: For further guidance on purge ventilation, see BS 5925:1991, *Code of Practice for Ventilation Principles and Designing for Natural Ventilation.*

The various methods of ventilation are summarized in Table 8.1.

Table 8.1 **Ventilation methods**

Method	Type	Why used	Remarks
Extract ventilation	Intermittent extract fans	In rooms where most water vapour and/or pollutants are released (e.g. cooking, bathing or photocopying)	May be either intermittent or continuous and is aimed at minimizing the spread of vapour and pollutants to the rest of the building
Whole building ventilation	Trickle ventilators	To provide fresh air to the building, dilute and disperse residual water vapour and pollutants not dealt with by extract ventilation, and to remove water vapour exchange with a ventilation rate that can be reduced or ceased when the building is not occupied. In some cases (e.g. when the building is reoccupied) it may be necessary to purge the air (see below)	Provides continuous air to remove pollutants released by building materials, furnishings, activities and the presence of occupants
Purge ventilation	Windows	To assist in the removal of high concentrations of pollutants and water vapour released from occasional activities (e.g. painting and decorating)	Is intermittent and may be used to improve thermal comfort and/or overheating in summer (see Approved Documents L1A (New Dwellings) and L2A (New Buildings))

8.2 Requirements

There shall be adequate means of ventilation provided for people in the building.   BR-AP-F

8.3 Meeting the requirements

Ventilation of a building is required to prevent the accumulation of moisture that could eventually lead to mould growth and pollutants originating from within the building, which could present a risk to the health of the occupants. To overcome this situation, every building must be designed and constructed in such a way that the air quality inside the building is not a threat to the health of the occupants or the capability of the building to resist moisture, decay or infestation.

A building should have provision for ventilation by either:	SBR-D-3.14.1
• natural means;	SBR-ND-3.14.1
• mechanical means; or	
• a combination of natural and mechanical means.	
Ventilation should have the capability of:	SBR-D-3.14.1
• rapidly diluting water vapour, where necessary, that is produced in apartments and sanitary accommodation;	SBR-ND-3.14.1
• removing excess water vapour from areas where it is produced in significant quantities (such as kitchens, utility rooms, bathrooms and shower rooms) to reduce the likelihood of creating conditions that support the germination and growth of mould, harmful bacteria, pathogens and allergens;	
• providing outside air to maintain indoor air quality sufficient for human respiration.	
Ventilation should always be to the outside (i.e. external) air.	SBR-D-3.14.1
	SBR-ND-3.14.1

There is no need to ventilate:

- a room (except for a kitchen or utility room) with a floor area of not more than $4\,m^2$;
- a store room that is only used for storage, provided that it is temperature controlled.

9

Floors

The ground floor of a brick-built house is either solid concrete or a suspended timber type. With a concrete floor, a damp-proof membrane (DPM) is laid between walls. With timber floors, sleeper walls of honeycomb brickwork are built on over site-concrete between the base brickwork. A timber sleeper plate then rests on each wall and timber joists are supported on them. Their ends may be similarly supported, led into the brickwork or suspended on metal hangers. Floorboards are laid at right angles to joists and first floor joists are supported by the masonry or suspended on metal hangers.

Similar to a brick-built house, the floors in a timber-framed house are either solid concrete or suspended timber. In some cases, a concrete floor may be screeded or surfaced with timber or chipboard flooring. Suspended timber floor joists are then supported on wall plates and surfaced with chipboard.

9.1 Requirements

The floors of the building shall adequately protect the building and people who use the building from harmful effects caused by the spillage of water from or associated with sanitary fittings or fixed appliances.

9.1.1 Conservation of fuel and power

Reasonable provision shall be made for the conservation of fuel and power in buildings by/from pipes, ducts and vessels used for hot water services.

9.1.2 Protection against sound within a dwelling-house, etc.

Dwelling-houses, flats and rooms for residential purposes shall be designed and constructed so that internal walls between a bedroom and a room containing a water closet shall provide reasonable resistance to sound.

9.2 Meeting the requirements

9.2.1 General

Floors next to the ground should not be damaged by groundwater.	BR-AP-C4.2

9.2.2 Moisture resistance

In areas such as kitchens, utility rooms and bathrooms where water may be spilled, any board used as flooring, irrespective of the storey, should be moisture resistant.	BR-AP-C4.15

9.2.3 Ground-supported floors exposed to moisture from the ground

Unless subjected to water pressure, the ground of a ground-supported floor should be covered with dense concrete laid on a hardcore bed and a DPM as shown in Table 9.1.

9.2.4 Enclosures used for drainage and/or water supply pipes

Any enclosure designed for drainage and/or the supply of water should:

• be bounded by a compartment wall or floor, an outside wall, an intermediate floor, or a casing; • have internal surfaces (except framing members) of Class 0; • not have an access panel that opens into a circulation space or bedroom; • be used only for drainage, or water supply, or vent pipes for a drainage system.	BR-AP-B3 (11.8)
The casing to the enclosure should:	B3 7.6-7.9 (V1), B3 10.7 (V2)

- be imperforate except for an opening for a pipe or an access pane;
- not be of sheet metal;
- have (including the access panel) not less than 30 minutes' fire resistance.

The opening for a pipe, either in the structure or in the casing, should be as small as possible and fire-stopped around the pipe.

Table 9.1 Ground-supported floor construction

Hardcore	Should be well compacted and no greater than 600 mm deep. It should be of clean, broken brick or similar inert material, and free from materials including water-soluble sulphates in quantities that could damage the concrete
Concrete	Should be at least 100 mm thick to mix ST2 in BS 8500 or (if there is embedded reinforcement) to mix ST4 of BS 8500
Damp-proof membrane	May be above or below the concrete which is continuous with the damp proof courses in walls and piers, etc.
	If below the concrete, the membrane will be formed with a sheet of polyethylene, at least 300 mm thick with sealed joints and laid on a bed of material that will not damage the sheet
	If laid above the concrete, the membrane may be either polyethylene (but without the bedding material) or three coats of cold applied bitumen solution or similar moisture and water vapour-resisting material
	In each case it should be protected by either a screed or a floor finish, unless the membrane is pitchmastic or similar material that may also serve as a floor finish
Insulation	If placed beneath floor slabs, it should have sufficient strength to withstand the weight of the slab, the anticipated floor loading and any possible overloading during construction
	If placed below the damp-proof membrane, it should have low water absorption and (if considered necessary) should be resistant to contaminants in the ground
Timber floor finish	If laid directly on concrete, it may be bedded in a material that can also serve as a damp-proof membrane
	Timber fillets that are laid in the concrete as a fixing for floor finishes should be treated with an effective preservative unless they are above the damp-proof membrane

9.2.5 Sanitary appliances below flood level

The basements of approximately 500 buildings in Scotland are flooded each year when the sewers surcharge (i.e. the effluent runs back up the pipes because they are too small to take the required flow). For this reason (and especially in Scotland):

• wastewater from floor gullies below flood level should be drained by wastewater lifting plants or, where there is unlikely to be a risk to persons such as in a car park, via an anti-flooding device.	SBR-D-3.7.2, SBR-ND-3.7.2

9.2.6 Provision for washing down

Where communal solid waste storage is located within a building, such as where a refuse chute is utilized:

• the storage area should have provision for washing down and draining the floor into a wastewater drainage system; • floor gullies should incorporate a trap that maintains a seal even during periods of disuse; • floors should be of an impervious surface that can be washed down easily and hygienically.	SBR-D-3.25.4

9.2.7 Hot water underfloor heating

All heat pumps (warm and/or hot water) are at their most efficient when the source temperature is as high as possible, the heat distribution temperature is as low as possible and pressure losses are kept to a minimum.

Supply water temperatures should be in the range of 30–40°C to an underfloor heating system.	SBR-D-6.3.4
The following controls should be fitted to ensure safe system operating temperatures: • a separate flow temperature high limit thermostat for warm water systems connected to any high water temperature heat supply; and • a separate means of reducing the water temperature to the underfloor heating system.	SBR-D-6.3.8

For hot water systems, zone controls are not considered necessary for single apartment dwellings.

Each zone (not exceeding 150 m^2) should have a room thermostat and a single multi-channel programmer or multiple heating zone programmers.   SBR-D-6.3.9
For large dwellings (with a floor area over 150 m^2) independent time and temperature control of multiple space heating zones is recommended.

10

Roofs

The roof of a building shall adequately protect the building and people who use the building from harmful effects caused by precipitation.

(BR-AP-C2)

10.1 Requirements

Gutters and rainwater pipes may be omitted from a roof at any height provided it has an area of not more than 8 m^2 and no other area drains onto it.

10.2 Meeting the requirements

10.2.1 Precipitation

Every building must be designed and constructed:	SBR-D-3.10, SBR-ND-3.10
• so that there will be no threat to the building or the health of the occupants as a result of moisture from precipitation penetrating to the inner face of the building;	
• so that the air quality inside the building is not a threat to the capability of the building being able to resist moisture, decay or infestation;	SBR-D-3.14, SBR-ND-3.14

• with a surface water drainage system that will:
  ○ ensure the disposal of surface water;
  ○ remove rainwater from the roof, or other areas where rainwater might accumulate, without causing damage to the structure.
	SBR-D-3.6, SBR-ND-3.6
	SBR-D-3.6.1, SBR-ND-3.6.1

Note: Roofs with copper, lead, zinc and other sheet metal roof coverings require provision for expansion and contraction of the sheet material. In 'warm deck' roofs, in order to reduce the risk of condensation and corrosion, it may be necessary to provide a ventilated air space on the cold side of the insulation and a high-performance vapour control layer between the insulation and the roof structure.

10.2.2 Flat roof constructions

See Table 10.1.

Table 10.1 **Roof constructions (flat)**

Type	Subtype		Common requirements	System-specific requirements
Flat roof constructions	Roof type A (concrete – warm roof)	Weatherproof covering / Insulation / Vapour control layer / Screed, if required	Flat roof structure of *in situ* or precast concrete with or without a screed; with or without a ceiling or soffit	External weatherproof covering. Insulation laid on a vapour-control layer between the roof structure and the weatherproof covering [a]
	Roof type B (concrete – inverted roof)	Protective covering / Insulation / Weatherproof covering / Screed, if required		External protective covering, with low-permeability insulation laid on a waterproof membrane between the roof structure and the external covering
	Roof type C (timber or metal frame – warm roof)	Weatherproof covering / Insulation / Vapour control layer	Flat roof structure of timber or metal-framed construction with a board decking 19 mm thick; with or without a ceiling or soffit	External weatherproof covering. Insulation laid on a vapour-control layer between the roof structure and the weatherproof covering [a]

(Continued)

Table 10.1 Continued

Type	Subtype	Common requirements	System-specific requirements
	Roof type D (timber or metal frame – inverted roof)		External protective covering, with low-permeability insulation laid on a waterproof membrane between the roof structure and the external covering
	Roof type E (troughed metal decking – warm roof)	Flat roof structure of timber or metal framed construction with a troughed metal decking; with or without a ceiling or soffit	External weatherproof covering. Insulation laid on a vapour-control layer between the roof structure and the weatherproof covering [a]
	Roof type F (troughed metal decking – inverted roof)		External protective covering, with low-permeability insulation laid on a waterproof membrane between the roof structure and the external covering

- Protective covering
- Insulation
- Weatherproof membrane

[a] Roof types A, C and E are not suitable for sheet metal coverings that require joints to allow for thermal movement.

11

Chimneys and Fireplaces

11.1 Requirements

Unless there are adequate damp-proof courses and flashings, etc., materials in chimneys can collect rainwater and deliver it to other parts of the dwelling below roof level.

A roof exposed to precipitation, or wind-driven moisture, should prevent penetration of moisture to the inner surface of any part of a building.	SBR-D-3.10.1, SBR-ND-3.10.1

11.2 Meeting the requirements

11.2.1 Flashing

Flashing is thin continuous piece or pieces of sheet metal (or other impervious material) installed to prevent the passage of water into a structure from an angle or a joint (Figure 11.1). Flashing generally operates on the principle that, in order for the water to penetrate a joint, it must first work itself upwards against the force of gravity or wind-driven rain. In both instances the water would have to follow a tortuous path during which the driving force will be dissipated. Exterior building materials can also be configured with a non-continuous profile to defeat water surface tension.

External flashing is usually made out of a sheet metal, such as aluminium, copper, painted galvanized steel, stainless steel, zinc alloy, terne metal, lead or lead-coated copper. Metal flashing should be provided with expansion joints on long runs to prevent deformation of the metal sheets and the selected metal should not stain or be stained by adjacent materials or react chemically with them.

 Note: Roofs with copper, lead, zinc and other sheet metal roof coverings require provision for expansion and contraction of the sheet material.

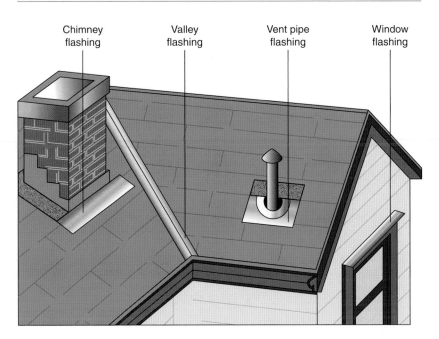

| Chimney flashing | Valley flashing | Vent pipe flashing | Window flashing |

Figure 11.1 Flashing on a roof

Flashing that is concealed within a construction assembly can be a sheet of metal or a waterproofing membrane such as bituminous fabric or plastic sheet material, depending on the climate and structural requirements. Aluminium and lead react chemically with cement mortar and are not normally used.

Roof flashing is placed around discontinuities or objects that protrude from the roof of a building (such as pipes and chimneys, or the edges of other roofs) to deflect water away from seams or joints.

Wall flashing may be embedded in a wall to direct water that has penetrated the wall back outside, or it may be applied in a manner intended to prevent the entry of water into the wall. Wall flashing is typically found at interruptions in the wall, such as windows and points of structural support.

Sill flashing is a concealed flashing that is typically placed under windowsills or door thresholds to prevent water from entering a wall at those points.

Base flashing is normally found at the base of a wall, and usually incorporates through-wall flashing with weep holes to permit the escape of water. Base flashings may be placed at the building grade or at a point where a roof intersects a wall.

The name flashing may derive from the fact that metal flashing (typically copper or aluminium) reflects flashes of sunlight. Historically, lead was also used for flashing, but this caused health and environmental problems and so, nowadays, rubber and other waterproof synthetic materials are used.

12

Sanitary Facilities

12.1 Requirements

Note: The following mandatory requirements have been extracted from the various Building and Water Regulations.

12.1.1 Sanitary facilities

All dwellings (houses, flats or maisonettes) should have at least one closet and one washbasin which should, ideally, be located in the room containing the closet. Sanitary facilities should:

- have smooth, non-absorbent surfaces and be capable of being easily cleaned;
- be capable of being flushed effectively;
- only be connected to a flush pipe or discharge pipe;
- have a regular supply of hot and cold water to the washbasin;
- be separated by a door from any space used for food preparation or where washing-up is done in washbasins.

(Approved Document G1)

Under existing regulations, all buildings (except factories and buildings used as workplaces) shall be provided with sufficient closet accommodation according to the intended use of that building and the amount of people using that building. The only exceptions are if the building (in the view of the local authority) has an insufficient water supply and a sewer is not available.

If a building already has sufficient water supply and the availability of a sewer, then the local authority has the authority to insist that the owner of the property replaces any other closet (e.g. an earth closet) with a water closet (WC). In these cases the owner is entitled to claim 50 per cent of the expense of doing this from the local authority.

If the local authority completes the work, then they are entitled to claim 50 per cent back from the owner.

The owner of the property has no right of appeal in these cases.

To meet this ruling, every building must be designed and constructed in such a way that:

- sanitary facilities are provided for all occupants of, and visitors to, the building in a form that allows convenience of use and that there is no threat to the health and safety of occupants or visitors;
- the products of combustion are carried safely to the external air without harm to the health of any person through leakage, spillage or exhaust, nor permit the re-entry of dangerous gases from the combustion process of fuels into the building.

12.1.2 Water closets

All plans for buildings must include at least one (or more) water or earth closet unless the local authority is satisfied in the case of a particular building that one is not required (for example in a large garage separated from the house).

If you propose using an earth closet, the local authority cannot reject the plans unless they consider that there is an insufficient water supply to that earth closet.

Every closet should:

- have a wash hand basin either within the toilet itself or in an adjacent space providing the sole means of access to the toilet; and
- be separated by a door from any room or space used wholly or partly for the preparation or consumption of food; and
- not open directly on to any room or space used wholly or partly for the preparation or consumption of food on a commercial basis; and
- include at least one enlarged WC cubicle where sanitary accommodation contains four or more WC cubicles in a range.
- be capable of being flushed effectively;
- only be connected to a flush pipe or discharge pipe;
- have smooth, non-absorbent surfaces and be capable of being easily cleaned.

Note: In the Greater London area, a 'water closet' can also be taken to mean a urinal.

Each closet should have one washbasin which should:

- ideally, be located in the room containing the closet;
- have a supply of hot and cold water.

Business premises

There may be some additional requirements regarding numbers, types and siting of appliances in business premises. If this applies to you then you will need to look at:

- the Offices, Shops and Railway Premises Act 1963;
- the Factories Act 1961; or
- the Food Hygiene (General) Regulations 1970.

12.1.3 Urinals

Every urinal that is cleared by water after use shall be supplied with water from a flushing device which:

- in the case of a flushing cistern, is filled at a rate suitable for the installation;
- in all cases, is designed or adapted to supply no more water than is necessary for effective flow over the internal surface of the urinal and for replacement of the fluid in the trap.

Note: The requirement is deemed to be satisfied:

- in the case of an automatically operated flushing cistern servicing urinals which is filled with water at a rate not exceeding:
  - ○ 10 litres per hour for a cistern serving a single urinal;
  - ○ 7.5 litres per hour per urinal bowl or stall, or, as the case may be, for each 700 mm width of urinal slab, for a cistern serving two or more urinals;
- in the case of a manually or automatically operated pressure flushing valve used for flushing urinals which delivers not more than 1.5 litres per bowl or position each time the device is operated.

12.1.4 Bathrooms and shower facilities

All dwellings (whether they are a house, flat or maisonette) should have at least one bathroom with a fixed bath or shower, and the bath or shower should be equipped with hot and cold water. This ruling applies to all plans for:

- new houses;
- new buildings, part of which are going to be used as a dwelling;
- existing buildings that are going to be converted, or partially converted into dwellings.

(Building Act 1984, section 27)

(BR-AP-G3)

Every building must be designed and constructed in such a way that protection is provided for people in, and around, the building from the danger of severe burns or scalds from the discharge of steam or hot water.

(BR-AP-G3)

12.1.5 Access and facilities for disabled people

Note: The aim of the amended Approved Document M to the Building Regulations is that suitable sanitary accommodation should be available to everybody, including sanitary accommodation specifically designed for wheelchair users, ambulant disabled people, people of either sex with babies and small children and/or people with luggage.

In addition to the requirements of the Disability Discrimination Act 1995, therefore, precautions shall be taken to ensure that:

- new non-domestic buildings and/or dwellings (e.g. houses and flats used for student living accommodation, etc.);
- extensions to existing non-domestic buildings;
- non-domestic buildings that have been subject to a material change of use (e.g. so that they become a hotel, boarding house, institution, public building or shop);

are capable of allowing people, regardless of their disability, age or gender, to use sanitary conveniences in the principal storey of any new dwelling.

(BR-AP-M)

If the proposed building is going to be used as a workplace or a factory in which persons of both sexes are going to be employed, then separate closet accommodation must be provided unless the local authority approves otherwise.

12.2 Meeting the requirements

12.2.1 Sanitary facilities

Every building must be designed and constructed in such a way that sanitary facilities are provided for all occupants of (and visitors to) the building in a form that allows convenience of use and that there is no threat to the health and safety of occupants or visitors.	SBR-D-3.12, SBR-D-3.12

In a residential building (other than a residential   SBR-D-2.9.16
care building and/or hospital) rooms intended for
use as sleeping accommodation (including any en
suite sanitary accommodation where provided)
should be separated from an escape route by a wall
providing short fire resistance duration.
Any door in the wall should be a suitable self-
closing fire door with a short fire resistance
duration.

Note: Although not recommending that sanitary facilities on the principal living level of a dwelling be designed to an optimum standard for wheelchair users (or any other disabled person), it should nevertheless be possible for most people to use these facilities unassisted and in privacy.

Number of sanitary facilities

The number of sanitary facilities provided within a building should be calculated from the maximum number of persons the building is likely to accommodate at any time, based on the normal use of the building.

Accessible toilets should be provided in accordance   SBR-ND-3.12.1
with the number of sanitary facilities recommended
in the various tables below and should be either:

- a minimum of one unisex accessible toilet,
  accessed independently from any other sanitary
  accommodation; or (if provided within separate
  sanitary accommodation for males and females)
- there should be at least one accessible toilet for
  each sex.

The number required will be dependent on travel distances within a building to an accessible toilet.

Separate male and female sanitary accommodation is   SBR-ND-3.12.1
usually provided based on the proportion of males and
females that will use a building or
(if the proportion is unknown) for equal numbers
of each sex.

Unisex sanitary accommodation may be provided for   SBR-ND-3.12.1
each sanitary facility (e.g. a WC and wash hand basin)
as long as it is located within a separate space, and is
intended to be used by only one person at a time, and
can be secured with a door from within so as to provide
privacy.

In small premises, it is recognized that duplication of sanitary facilities may not always be reasonably practicable and that they may be shared between staff and customers. However, where possible, it is always good practice that sanitary facilities for staff involved in the preparation or serving of food or drink are reserved for their sole use, with a separate provision made for customers. Separate hand-washing facilities for such staff should *always* be provided.

In Scotland the numbers of sanitary facilities   SBR-ND-3.12.1
in schools should be provided in accordance
with the tables in the School Premises (General
Requirements and Standards) (Scotland)
Regulations 1967, as amended.
Residential care buildings, day care centres and   SBR-ND-3.12.1
hospices may be subject to additional standards
set out in the relevant National Care Standards
document for that service.

Sanitary accommodation: dwellings

A dwelling should have at least one accessible WC, or waterless closet, and wash hand basin and at least one accessible shower or bath.

Bathrooms and toilets designed to minimum space standards can often create difficulties in use. As the ability of occupants can vary significantly, sanitary accommodation should both be immediately accessible and offer potential for simple alteration in the future.

To ensure that privacy can be maintained, if there is only one sanitary accommodation in a dwelling then it should not be en suite, nor reached through such an apartment.

Sanitary facilities should be: SBR-D-3.12.3

• located on the principal living level of a dwelling;

- of a size and form that allows unassisted use, in privacy, by almost any occupant;
- capable of being used by a person with mobility impairment or who uses a wheelchair.

Accessible sanitary accommodation should have: SBR-D-3.12.3

- a manoeuvring space (at least 1.1 m long by 800 mm wide, oriented in the direction of entry, and clear of any door swing or other obstruction) that will allow a person to enter and close the door behind them;
- unobstructed access at least 800 mm wide to each sanitary facility;
- an activity space for each sanitary facility, as noted in Figure 12.1;
- an unobstructed height above each activity space and above any bath or shower of at least 1.8 m above floor level;
- walls adjacent to any sanitary facility that are of robust construction that will permit secure fixing of grab rails or other aids; and
- where incorporating a WC, space for at least one recognized form of unassisted transfer from a wheelchair to the WC.

Notes:

- The activity space in front of a WC need not be parallel with the axis of the WC.
- Where allowing for side transfer, a small wall-hung wash hand basin may project up to 300 mm into the activity space in front of the WC.
- The projecting rim of a wash hand basin may reduce the width of a route to another sanitary facility to not less than 700 mm.
- A hand-rinse basin should only be installed within a toilet if there is a full-size wash hand basin elsewhere in the dwelling.

An additional accessible toilet may be needed on the entrance level of a dwelling where this is not also the principal living level.	SBR-D-3.12.3
An accessible shower room should be of a size that will accommodate both a level-access floor shower with a drained area of not less than 1.0 m × 1.0 m (or equivalent) or a 900 mm × 900 mm shower tray (or equivalent).	SBR-D-3.12.3

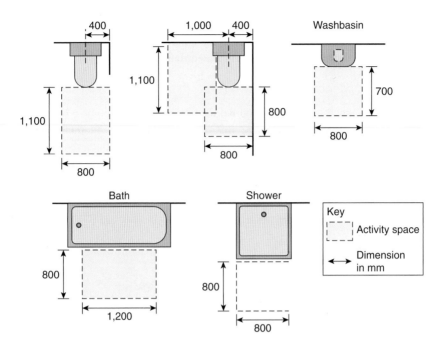

Figure 12.1 Activity spaces for accessible sanitary facilities

An accessible bathroom should be of a size that will accommodate a 1.7 m × 700 mm bath (or equivalent).	SBR-D-3.12.3
Within an accessible bathroom, it should be possible to replace the bath with an accessible shower without adversely affecting access to other sanitary facilities.	SBR-D-3.12.3
The activity space in front of a bath may be at any position along its length.	SBR-D-3.12.3

If a dwelling has a bathroom or shower room on another level (which is not en suite to a bedroom) some occupants may not require the immediate provision for bathing on the principal living level. If this is the case, then the principal living level may then have a separate, enclosed space that is capable of incorporating an accessible toilet and can be sufficient to incorporate an accessible shower room (as described above) in the future.

This space should nevertheless have a drainage connection that is positioned so as to allow the installation of a floor shower or raised shower tray, sealed and terminated either immediately beneath floor level, under a removable access panel, or at floor level in a visible position. The structure and insulation of the floor in any area identified for a future floor shower should allow for the depth of an inset tray installation or, if it is a solid floor, have a 'laid to fall' installation. If not adjacent to an accessible toilet and separated by an easily demountable partition, a duct to the external air should be provided to allow for later installation of mechanical ventilation.

Sanitary appliances below flood level

Sanitary appliances that are located below flood level can become a particular problem. As previously mentioned, this occurs frequently in Scotland, where many basements are flooded each year when the sewers surcharge and the effluent runs back up the pipes because they are too small to take the required flow.

Wastewater from sanitary appliances and floor gullies below flood level should be drained by wastewater lifting plants (constructed in accordance with the requirements of BS EN 12056-4:2000) or, where there is unlikely to be a risk to persons such as in a car park, via an anti-flooding device.	SBR-D-3.7.2, SBR-ND-3.7.2
Wastewater from sanitary appliances above flood level should not be drained through anti-flooding devices and only in special cases (e.g. refurbishment) by a wastewater lifting plant.	SBR-D-3.7.2, SBR-ND-3.7.2

Loan of temporary sanitary conveniences

If the local authority is maintaining, improving or repairing drainage systems and this requires the disconnection of existing buildings from these sanitary conveniences, then, on request from the occupier of the building, the local authority is required to supply (on temporary loan and at no charge) sanitary conveniences:

- if the disconnection is caused by a defect in a public sewer;
- if the local authority has ordered the replacement of earth closets (see above);
- for the first seven days of any disconnection.

Sanitary pipework

Sanitary pipe work should be constructed and installed in accordance with the recommendations in BS EN 12056-2:2000 and according to the type of system used as shown in Table 12.1.	SBR-D-3.7.1, SBR-ND-3.7.1

Table 12.1 Sanitary discharge stack systems

System I	Single discharge stack system with partially filled branch discharge pipes
System II	Single discharge stack system with small bore discharge branch
System III	Single discharge stack system with full bore branch discharge pipes (the traditional system in use in the UK)
System IV	Separate discharge stack system (mainly found in various continental European countries and unlikely to be appropriate for use in Scotland)

Protection of openings for pipes

Pipes that pass through a compartment wall or compartment floor (unless the pipe is in a protected shaft), or through a cavity barrier, should conform to one of the following alternatives:

Proprietary seals – that maintain the fire resistance of the wall, floor or cavity barrier (any pipe diameter).	B3 (11.5 –11.6)
Pipes – with a restricted diameter where fire-stopping is used around the pipe, keeping the opening as small as possible.	B3 (11.5 and 11.7
Sleeving – a pipe of lead, aluminium, aluminium alloy, fibre-cement or UPVC, with a maximum nominal internal diameter of 160 mm, may be used with a sleeving of non-combustible pipe as shown in Figure 12.2.	B3 (11.5 and 11.8)

Note: The opening in the structure should always be made as small as possible so as to provide fire-stopping between pipe and structure.

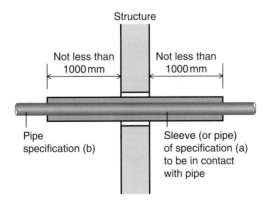

Figure 12.2 Pipes penetrating a structure

Ventilation of sanitary accommodation

Every building must be designed and constructed in such a way that the air quality inside the building is not a threat to the health of the occupants or the capability of the building to resist moisture, decay or infestation.

Extract ventilation (i.e. the removal of air directly from a space or spaces to outside) may be by natural means such as passive stack ventilation (PSV) or by mechanical means such as by an extract fan or a central system.

All sanitary accommodation shall be provided with extract ventilation to the outside which is capable of operating either intermittently or continuously with a minimum extract rate of 6 litres/second.	SBR-ND-3.14.1, SBR-D-3.14.1, BR-AP-F 1.5
Ventilation devices designed to work continuously shall not have automatic controls such as a humidity control when used for sanitary accommodation.	BR-AP-F Table 1.5
As odour is the main pollutant, humidity controls should not be used for intermittent extract in sanitary accommodation.	BR-AP-F Table 1.5
Ventilation should be to the outside (i.e. external) air.	SBR-D-3.14.1, SBR-ND-3.14.1
PSV can be used as an alternative to a mechanical extract fan for office sanitary and washrooms and food preparation areas (BS 8300).	BR-AP-F Table 2.2a
PSV devices shall have a minimum internal duct diameter of 80 mm and a minimum internal cross-sectional area of 5000 mm^2.	BR-AP-F-App D

Note: Purge ventilation may be used provided that security is not an issue.

Common outlet terminals and/or branched ducts shall not be used for wet rooms such as WCs.	BR-AP-F-App D
Where there was no previous ventilation opening, or where the size of the original ventilation opening is not known, replacement window(s) shall have an equivalent area greater than 2500 mm^2 per WC.	BR-AP-F-3.6
In buildings other than dwellings, fresh air supplies should be protected from contaminants that could be injurious to health.	BR-AP-F-2.3
All office sanitary accommodation and washrooms shall be provided with intermittent air extract.	BR-AP-F-2.11

Ventilation of sanitary accommodation in non-domestic buildings

Any area containing sanitary facilities should be well ventilated, so that offensive odours do not linger.	SBR-ND-3.14.7
Measures should be taken to prevent odours entering other rooms (e.g. by providing a ventilated area between the sanitary accommodation and the other room or by using mechanical ventilation).	SBR-ND-3.14.7
No room containing sanitary facilities should communicate directly with a room for the preparation or consumption of food.	SBR-ND-3.14.7

If a conservatory and/or an extension is built over an existing window and/or ventilators, then, if is constructed over an area that generates moisture, such as a bathroom or shower room:

Mechanical extract, via a duct or a PSV system, should be provided direct to the outside air.

Sanitary facilities: non-domestic buildings

To ensure safety, ease of use and hygiene, the following provisions should be made within *all* sanitary accommodation:

• sanitary facilities, fittings and surface finishes should be easily cleanable, so as to ensure that a hygienic environment is maintained;	SBR-ND-3.12.6
• where a door opens into a space containing a sanitary facility, there should be an unobstructed space of at least 450 mm in diameter between the sanitary facility and the door swing (see Figure 12.3); and	
• if a door is fitted with a privacy lock, then it should have an emergency release, that is able to be operated from the outside and (if not sliding or opening outward) offer an alternate means of removal, to permit access in an emergency;	
• all sanitary facilities (and any associated aid or fitting, such as a grab rail, etc.), should contrast visually with surrounding surfaces so as to assist people with visual impairments.	
A clear glazed vision panel should be provided to the outer door of a lobby that only provides access (i.e. solely) to sanitary accommodation.	SBR-ND-4.2.5

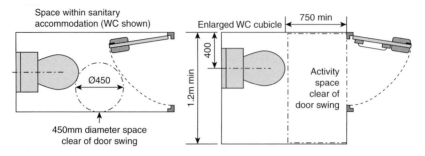

Figure 12.3 Space within sanitary accommodation and enlarged WC cubicle

Provision in residential buildings

Public expectation of facilities in residential buildings has risen considerably over the years and en suite sanitary facilities now tend to be normal practice, although it is recognized that this may not be possible in all cases.

> Where sanitary accommodation is not en suite   SBR-ND-3.12.3
> to bedrooms, it should be located directly off a
> circulation area, close to bedrooms and provided in
> accordance with Table 12.2.

Sanitary accommodation containing a bath or shower should also contain a WC and a wash hand basin, in addition to the general provision for those sanitary facilities noted in Table 12.2.

Table 12.2 Number of sanitary facilities in residential buildings

Sanitary facility	Number of sanitary facilities
WC	1 per 9 persons, or part thereof
WHB	1 per bedroom
Bath or shower	1 per 4 persons, or part thereof

For bedrooms, a wash hand basin should be en suite. An alternate ratio of one wash hand basin per four persons, or part thereof, may be used for dormitory sleeping accommodation.

> In a residential building, an accessible bedroom should be   SBR-ND-
> provided with sanitary accommodation   3.12.3

(comprising a WC, wash hand basin and a bath or shower) which should be en suite, other than:

- when altering or converting an existing building, it is not reasonably practicable to provide en suite sanitary accommodation; or
- where sanitary facilities need to be kept separate for safety reasons, such as in a place of lawful detention.

Where accessible sanitary accommodation is not en suite, it should be located directly off a circulation area, close to any accessible bedroom, and should be clearly identified.

SBR-ND-3.12.3

Public conveniences

You are not allowed to erect a public sanitary convenience in (or on) any location that is accessible from a street, without the consent of the local authority. Any person who contravenes this requirement is liable to a fine and can be made (at his own expense) to remove or permanently close it.

Note: This requirement does not apply to sanitary conveniences erected by a railway company within their railway station, yard and/or approaches, and by dock undertakers on land belonging to them.

Workplace conveniences

If the building is a workplace used by both sexes, then sufficient and satisfactory accommodation is required for persons of each sex.

This requirement does not apply to premises to which the Offices, Shops and Railway Premises Act of 1963 applies.

Sanitary provision for staff

A building should be provided with sanitary facilities for staff in accordance with Table 12.3.

SBR-ND-3.12.2

Table 12.3 **Number of sanitary facilities for staff**

	Staff numbers	WCs	WHBs	Urinals
Male	1–15	1	1	1
	16–30	2	2	1
	31–45	2	2	2
	46–60	3	3	2

(Continued)

Table 12.3 Continued

	Staff numbers	WCs	WHBs	Urinals
	61–75	3	3	3
	76–90	4	4	3
	91–100	4	4	4
	>100	1 additional WC, WHB and urinal for every additional 50 males or part thereof		
Female (also male where no urinals are provided)	1–5	1		
	6–25	2	1	
	>25	1 additional WC, WHB and urinal for every additional 25 females (or males) or part thereof		

Provision for the public in shops and shopping malls

Sanitary accommodation for customers within shops   SBR-ND-3.12.4
and shopping malls should be:

- clearly identified and located so that it may be easily reached;
- provided on the entrance storey and, in larger buildings of more than two storeys, with a total sales floor area greater than $4000\,m^2$, on every alternate storey;
- in accordance with Table 12.4.

Table 12.4 Number of sanitary facilities for people, other than staff, in shops

Building type		Sales area of shop (m^2)	WCs	Urinals
Shops (class 1) and shopping malls	Unisex	500–1000	1	
		1001–2000	1	1
		2001–4000	1	2
		>4000	Plus 1 WC for each extra $2000\,m^2$ of sales area, or part thereof	Plus 1 urinal for each extra $2000\,m^2$ of sales area, or part thereof
	Female	1000–2000	2	
		2001–4000	5	
		>4000	Plus 2 WCs for each extra $2000\,m^2$ of sales area, or part thereof	

(Continued)

Table 12.4 **Continued**

Building type		Sales area of shop (m²)	WCs	Urinals
Shops (class 2) and shopping malls	Male	1000–2000	1	1
	Male	>4000	1, plus 1 WC for each extra 3000 m² of sales area, or part thereof	1, plus 1 urinal for each extra 3000 m² of sales area, or part thereof
	Female	1000–2000	3	
		2001–4000	1, plus 1 WC for each extra 3000 m² of sales area, or part thereof	
		>4000		

One wash hand basin should be provided for each WC, plus one wash hand basin per five urinals, or part thereof.

Notes: For the purposes of this guidance, shop sales areas are classified as:

- Class 1 supermarkets and department stores (all sales areas), plus:
    - shops for personal services such as hairdressing;
    - shops for the delivery or uplift of goods for cleaning, repair or other treatment or for members of the public themselves carrying out such cleaning, repair or other treatment;
- Class 2 shop sales areas in shops trading predominantly in furniture, floor coverings, bicycles, perambulators, large domestic appliances or other bulky goods, or trading on a wholesale self-selection basis.

For shopping malls, the sum of the sales areas of all the shops in the mall should be calculated and used with this table. Sanitary facilities provided within a shop may be included in the overall calculation.	SBR-ND-3.12.4
If a shop has a restaurant or café, additional sanitary facilities to serve the restaurant should be provided.	SBR-ND-3.12.4

Public conveniences in entertainment and assembly buildings

Provision of sanitary facilities for public in entertainment and assembly buildings should be in accordance with Table 12.5.	SBR-ND-3.12.5
In cinema multiplexes and similar premises (where the use of sanitary facilities will be spread through the opening hours) the level of sanitary facilities should normally be based upon 75% of total capacity.	SBR-ND-3.12.5

Table 12.5 Number of sanitary facilities for entertainment and assembly buildings

Building type		No. of people	WCs	Urinals
Buildings used for assembly or entertainment (e.g. places of worship, libraries, cinemas, theatres, concert halls and premises without licensed bars)	Male	1–100	1	2
		101–250	1	Plus 1 for each
		>250	Plus 1 for each extra 500 males or part thereof	extra 50 males or part thereof over 100
	Female	1–40	3	
		40–70	4	
		71–100	5	
		>100	Plus 1 for each extra 500 males or part thereof	
Restaurants, cafés, canteens and fast food outlets (where seating is provided)	Male	1–400	1 for every 100 or part thereof, plus 1 for each extra 250 males or part thereof	1 for 50 males or part thereof
		>400		
	Female	1–50	2	
		51–100	3	
		101–150	4	
		151–200	5	
		>200	6, plus 1 for each extra 100 females, or part thereof	
Public houses and licensed bars	Male	1–75	1	2
		76–150	1	3
		>150	Plus 1 for each extra 150 males, or part thereof	Plus 1 for each extra 75 males or part thereof
	Female	1–10	1	
		11–25	2	
		>25	Plus 1 for each extra 25 females or part thereof	
Swimming pools (bathers only)	Male	1–100	2	1 per 20 males
		>100	Plus 1 for each extra 150 males, or part thereof	
	Female	1–25	2	
		>25	2, plus 1 for each extra 25 females or part thereof	

Note: For single-screen cinemas, 100 per cent occupancy is assumed.

For works and office canteens, the amount of sanitary facilities may be reduced proportionally if there are   SBR-ND-3.12.5

readily accessible workplace sanitary facilities close to
the canteen.

In public houses, the number of customers should be calculated at the rate of four persons per 3 m^2 of effective drinking area (i.e. the total space of those parts of those rooms to which the public has access).	SBR-ND-3.12.5
Public houses with restaurants and licences for public music, singing and dancing should be provided with sanitary facilities as for licensed bars.	SBR-ND-3.12.5
Sanitary facilities intended for spectators should be provided in accordance with buildings used for public entertainment.	SBR-ND-3.12.5

Within residential buildings or sports facilities (where bathing or showering forms an integral part of the activities performed there) a person should be able to use sanitary facilities in privacy, with or without assistance.

Hot water discharge from sanitary fittings

To prevent the development of Legionella or similar pathogens, hot water within a storage vessel should be stored at a temperature of not less than 60°C and distributed at a temperature of not less than 55°C.	SBR-D-4.9.5, SBR-ND-4.9.5
To prevent scalding, the temperature of hot water, at point of delivery to a bath or bidet, should not exceed 48°C.	SBR-D-4.9.5, SBR-ND-4.9.5
A device or system limiting water temperature should allow flexibility in setting of a delivery temperature, up to a maximum of 48°C, in a form that is not easily altered by building users.	SBR-D-4.9.5, SBR-ND-4.9.5

12.2.2 Water closets

Every water closet, and every flushing device designed for use with a water closet, shall comply with a specification approved by the Regulator for the purposes of this Schedule.	SI 1148-25(2)

The requirements of sub-paragraphs 25(1) and 25(2) do not apply where faeces or urine are disposed of through an appliance that does not solely use fluid to remove the contents.

Every water closet pan shall be supplied with water from a   SI 1148
flushing cistern, pressure flushing cistern or pressure flushing   25(1)(a)
valve, and shall be so made and installed that after normal
use its contents can be cleared effectively by a single flush of
water, or, where the installation is designed to receive flushes
of different volumes, by the largest of those flushes.

No pressure flushing valve shall be installed: SI 1148
                                                   25(1)(b)
• in a house; or
• in any building not being a house; or
• where a minimum flow rate of 1.2 litres per second
   cannot be achieved at the appliance.

Where a pressure flushing valve is connected to a supply   SI 1148
pipe or distributing pipe, the flushing arrangement shall   25(1)(c)
incorporate a backflow prevention device consisting of a
permanently vented pipe interrupter located not less than
300 mm above the spillover level of the WC pan or urinal.

No flushing device installed for use with a WC   SI 1148 25(1)(d)
pan shall give a single flush exceeding 6 litres.

Note: Notwithstanding sub-paragraph 25(1)(d), a flushing cistern installed
before 1st July 1999 may be replaced by a cistern which delivers a similar
volume and which may be either single flush or dual flush, but a single flush
cistern may not be so replaced by a dual flush cistern.

No flushing device designed to give flushes of   SI 1148 25(1)(f),
different volumes shall have a lesser flush exceeding   SWB 25(1)(e)
two-thirds of the largest flush volume.

A flushing device designed to give flushes of   SI 1148 25(1)(h),
different volumes: SWB 25(1)(g)

• shall have a readily discernible method of
   actuating the flush at different volumes; and
• shall have instructions, clearly and
   permanently marked on the cistern or
   displayed nearby, for operating it to obtain the
   different volumes of flush.

Every flushing cistern, other than a pressure flushing cistern, shall be clearly marked internally with an indelible line to show the intended volume of flush, together with an indication of that volume.   SI 1148 25(1)(g), SWB 25(1)(f)

Every flushing cistern, not being a pressure flushing cistern or a urinal cistern, shall be fitted with a warning pipe or with a no less effective device.   SI 1148 25(1)(i), SWB 25(1)(h)

Water closets: dwellings

Every dwelling should have sanitary facilities comprising at least one water closet (WC), or waterless closet, together with one wash hand basin per WC, or waterless closet, one bath or shower and one sink.   BR-AP-G1, SBR-D-3.12.1

To allow for basic hygiene, a wash hand basin should always be close to a WC or waterless closet, either within a toilet, or located in an adjacent space providing the sole means of access to the toilet.   BR-AP-G1, SBR-D-3.12.1

There should be a door separating a space containing a WC, or waterless closet, from a room or space used for the preparation or consumption of food, such as a kitchen or dining room.   SBR-D-3.12.1

If a waterless closet is installed it should be to a safe and hygienic design.

The principal living level should be made accessible to as wide a range of occupants as possible and, accordingly:

- there should be an accessible toilet on the entrance level in addition to sanitary facilities on the principal living level;   SBR-D-4.2.6, SBR-D-4.2.10 BR-AP-G
- there should be level or ramped access from the entrance of the dwelling to this accessible toilet and at least one of the apartments on the entrance storey;
- closets (and/or urinals) should be separated by a door from any space used for food preparation or where washing-up is done;
- the surfaces of a closet, urinal or washbasin should be smooth, non-absorbent and capable of being easily cleaned;

- closets (and/or urinals) should be capable of being flushed effectively;
- closets (and/or urinals) should only be connected to a flush pipe or discharge pipe;
- closets and/or urinals fitted with flushing apparatus should discharge through a trap and discharge pipe into a discharge stack or a drain;
- closets fitted with macerators and pump can be connected to a discharge pipe discharging to a discharge stack if the macerator and pump system is approved under the current European Technical Approval system;
- washbasins should discharge through a grating, a trap and a branch discharge pipe to a discharge stack or (if it is a ground floor location) into a gully or directly into a drain;
- if there is no suitable water supply or means of disposing foul water, closets (and/or urinals) can use chemical treatment;
- although the above are the minimum requirements for meeting sanitary conveniences and washing facilities, other local authority regulations may apply and it is worth seeking the advice of the local planning officer before proceeding.

Water closets: non-domestic buildings

An accessible toilet should be provided in any building with toilet facilities.	SBR-ND-3.12.8
In large building complexes (e.g. retail parks and large sports centres) there should be one wheelchair-accessible unisex toilet capable of including an adult changing table.	BR-AP-M (5.17)
An accessible toilet should include a WC with:	SBR-ND-3.12.8

- a seat height of 480 mm, to assist in ease of transfer to and from a wheelchair; and
- a flush lever fitted to the transfer side of the cistern (Figure 12.4).

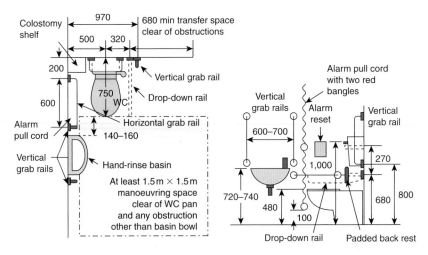

Figure 12.4 Provisions within an accessible toilet

Location of accessible toilets in shopping complexes

Accessible toilets should be located where they can be reached easily and the horizontal distance from any part of a building to an accessible toilet should be not more than 45 m.	SBR-ND-3.12.9
Where there are no toilets on a storey, all occupied parts of that storey should be within 45 m of the nearest accessible toilet on an adjacent storey.	SBR-ND-3.12.9
Within the retail area of a large superstore or the concourse of a shopping mall, the distance from an accessible toilet may be increased to not more than 100 m, provided there are no barriers, such as pass doors or changes of level on the route and the location of the accessible toilet is well signposted.	SBR-ND-3.12.9

Toilets and washrooms in protected zones

Provided all parts of the building served by an escape stair have at least one other escape route, toilet and washrooms not more than 3 m² may be located within a protected zone.	SBR-ND-2.9.23
The walls/doors separating the toilets or washrooms from the protected zone need not have a fire resistance duration.	SBR-ND-2.9.23

Ventilation of toilets

Most dwellings are naturally ventilated and ventilation should be provided in accordance with Table 12.6.

Table 12.6 **Recommended ventilation of a dwelling**

	Ventilation recommendations	Trickle ventilation $>10\,\text{m}^3/\text{h}/\text{m}^2$	Trickle ventilation $10\,\text{m}^3/\text{h}/\text{m}^2$
Toilet	Either: • a ventilator with an opening area of at least 1/30th of the floor area it serves; or • mechanical extraction capable of at least three air changes per hour.	4,000 mm²	10,000 mm²

Note: To reduce the effects of stratification of the air in a room, some part of the opening ventilator should be at least 1.75 m above floor level.

12.2.3 Urinals

Except in the case of a urinal which is flushed manually, or which is flushed automatically by electronic means after use, every pipe which supplies water to a flushing cistern or trough used for flushing a urinal shall be fitted with an isolating valve controlled by a time switch and a lockable isolating valve, or with some other equally effective automatic device for regulating the periods during which the cistern may fill.	SI 1148, 25(1)(k), SWB 25(1)(j)

12.2.4 Bathrooms and shower facilities

All dwellings (houses, flats or maisonettes) should have at least one bath-room with a fixed bath or shower. Bathrooms shall have either a fixed bath or shower bath that is provided with hot and cold water and connected to a foul water drainage system, and the bath or shower should:

• have a supply of hot and cold water;	BR-AP-G2
• discharge through a grating, a trap and branch discharge pipe to a discharge stack or (if on a ground floor) discharge into a gully or directly to a foul drain;	BR-AP-G2 and H1

- be connected to a macerator and pump (of an approved type) if there is no suitable water supply or means of disposing foul water.   BR-AP-G2

Every bathroom or shower room should have a floor surface that minimizes the risk of slipping when wet.   SBR-ND-3.12.6

Baths, sinks, showers and taps

All premises supplied with water for domestic purposes shall have at least one tap conveniently situated for the drawing of drinking water.   SI 1148-26

A drinking water tap shall be supplied with water from:
(a) a supply pipe;   SI 1148-27(a)
(b) a pump delivery pipe drawing water from a supply pipe; or   27(b)
(c) a distributing pipe drawing water exclusively from a storage cistern supplying wholesome water.   27(c)

Every bath, washbasin, sink or similar appliance shall be provided with a watertight and readily accessible plug or other device capable of closing the waste outlet.   SI 1148-28(1)

Note: This requirement does not apply to:

- an appliance where the only taps provided are spray taps;
- a washing trough or washbasin whose waste outlet is incapable of accepting a plug and to which water is delivered at a rate not exceeding 0.06 litres/second exclusively from a fitting designed or adapted for that purpose;
- a washbasin or washing trough fitted with self-closing taps;
- a shower bath or shower tray;
- a drinking water fountain or similar facility; or
- an appliance which is used in medical, dental or veterinary premises and is designed or adapted for use with an unplugged outlet.

Washbasins

Washbasins should have a supply of hot and cold water.   BR-AP-G1
Washbasins should discharge through a grating, a trap and a branch discharge pipe to a discharge stack or (if it is a ground floor location) into a gully or directly into a drain.   BR-AP-G1

For most people, a level access shower is generally both easier and more convenient to use than a bath and, therefore, should always be included within a building where sanitary facilities for bathing are provided.

Showers

An accessible shower (see Figure 12.5) should be separate or screened from other accommodation, to allow privacy when bathing.	SBR-ND-3.12.10

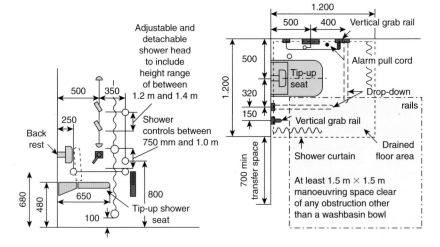

Figure 12.5 Provisions within an accessible shower room

A shower area without separating cubicles (such as found within sporting facilities, etc.) which comprises 10 or more showers should include at least one communal shower.	SBR-ND-3.12.10
To avoid undue waiting times, where an accessible bath or shower is combined with accessible toilet facilities, a separate accessible toilet should also be provided.	SBR-ND-3.12.10
If an accessible changing facility (see Figure 12.6) is combined with an accessible showering facility, a second fold-down seat should be fitted out within the showering area and manoeuvring space to assist in drying and changing.	SBR-ND-3.12.11

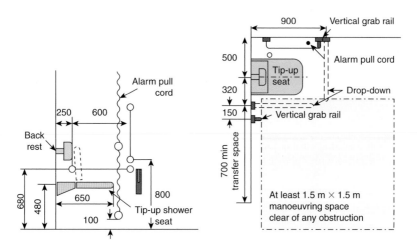

Figure 12.6 Provisions within an accessible changing facility

Oil-firing appliances in bathrooms and shower rooms

There is an increased risk of carbon monoxide poisoning if an oil-firing appliance is located in a bathroom and shower room. Because of this:

• open-flued oil-firing appliances should not be installed in these rooms or any cupboard or compartment connecting directly with these rooms;	SBR-D-3.20.5, SBR-ND-3.20.5
• where locating a combustion appliance in such rooms cannot be avoided, the installation of a room-sealed appliance would be appropriate.	

Gas-fired appliances in bathrooms and shower rooms

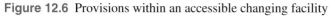

Regulation 30 of the Gas Safety (Installations and Use) Regulations 1998 has specific requirements for room-sealed appliances in these locations.

Ventilation of bathrooms and showers

Ventilation should have the capability of: • removing excess water vapour from bathrooms and shower rooms to reduce the likelihood of creating conditions that support	SBR-ND-3.14.1, SBR-D-3.14.1

the germination and growth of mould, harmful
bacteria, pathogens and allergens;
- rapidly diluting pollutants and water vapour,
  where necessary, that is produced in sanitary
  accommodation.

Most dwellings are naturally ventilated and ventilation should be provided
in accordance with Table 12.7.

Table 12.7 **Recommended ventilation of a dwelling**

	Ventilation recommendations	Trickle ventilation $>10\,m^3/h/m^2$	Trickle ventilation $10\,m^3/h/m^2$
Bathroom or shower-room (with or without a WC)	Either: • mechanical extraction capable of at least 15 litres/sec (intermittent); or • a passive stack ventilation system.	$4,000\,mm^2$	$10,000\,mm^2$

Note: To reduce the effects of stratification of the air in a room, some part of the opening
ventilator should be at least 1.75 m above floor level.

Natural ventilation

Where a building is naturally ventilated, all moisture-producing areas such as bathrooms and shower rooms should have the additional facility for removing such moisture before it can damage the building.	SBR-ND-3.14.2
Where rapid ventilation is provided (e.g. an opening window) some part of the opening should be at least 1.75 m above floor level.	SBR-ND-3.14.2

Trickle ventilators

Trickle ventilators should be provided in naturally ventilated areas to allow
fine control of air movement.

A trickle ventilator serving a bathroom or shower room may open into an area that does not generate moisture, such as a bedroom or hallway, provided the area is fitted with a trickle ventilator.	SBR-D-3.14.5

Electrical safety

Socket outlets in bathrooms and rooms containing a shower

In a bathroom or shower room, an electric shaver power outlet, complying with BS EN 60742:1996, may be installed. Other than this, there should be no socket outlets and no means for connecting portable equipment.	SBR-D-4.5.4, SBR-ND-4.5.4
Where a shower cubicle is located in a room, such as a bedroom, any socket outlet should be installed at least 3 m from the shower cubicle.	SBR-D-4.5.4, SBR-ND-4.5.4

Lighting

A dwelling should have an electric lighting system providing at least one lighting point to every bathroom with a floor area of 2 m² or more.	SBR-D-4.5.1

Access to manual controls

The location of any manual control device (i.e. for openable ventilators, windows and roof lights) and for controls and outlets of electrical fixtures located on a wall or other vertical surface that is intended for operation by the occupants of a building should be not more than 1.2 m above floor level within accessible sanitary accommodation not provided with mechanical ventilation.	SBR-D-4.8.5, SBR-ND-4.8.6

Fire detection and alarm systems

In hospitals, a fire detection and alarm system need not be provided in sanitary accommodation. When provided, however (e.g. in a disabled toilet), an assistance alarm should have an audible tone, distinguishable from any fire alarm, together with a visual indicator, both within the sanitary accommodation and outside in a location that will alert building occupants to the call.	SBR-ND-2.A.5, SBR-ND-2.B.5 SBR-ND-3.12.7

Controls for heating circuits

Bathrooms or en suites which share a heating circuit   SBR-D-6.3.8
with an adjacent bedroom:

- should provide heat only when the bedroom
  thermostat is activated;
- should be fitted with an independent towel rail
  or radiator.

Control of humidity

Control of moisture in bathrooms and shower rooms   SBR-D-3.15.2
can be by active or passive means.

12.2.5 Access and facilities for disabled people

Sanitary facilities

Note: All terminal sanitary fittings should comply with Guidance Note
G18.5 of the Guidance Document relating to Schedule 2: Requirements for
Water Fittings of the Water Supply (Water Fittings) Regulations 1999, SI
1999/1148.

Where sanitary facilities are provided in a building,   BR-AP-M (5.7b)
at least one wheelchair-accessible unisex toilet
should be available.

Accessible sanitary accommodation should   SBR-ND-3.12.7

- be clearly identified;
- contain a manoeuvring space of at least 1.5 m
  by 1.5 m, clear of any obstruction, including a
  door swing, other than a wall-mounted wash
  hand basin which may project not more than
  300 mm into this space;
- be fitted with fixed and folding grab rails (see
  Figure 12.7);
- be fitted with an assistance alarm which can be
  operated or reset when using a sanitary facility
  and which is also operable from floor level; and
- where more than one accessible sanitary facility
  of a type is provided within a building, offer
  both left- and right-hand transfer layouts.

The surface finish of sanitary fittings and grab   BR-AP-M (5.4k)
bars should contrast visually with background
wall and floor finishes.

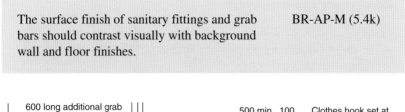

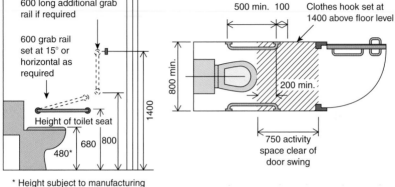

Figure 12.7 Example of a WC cubicle for an ambulant disabled person

Taps should be operable by people with limited   BR-AP-M (5.3)
strength and/or manual dexterity.
Bath and washbasin taps should either be controlled   BR-AP-M (5.4a)
automatically or be capable of being operated using
a closed fist, e.g. by lever action.

Doors

Doors to sanitary facilities intended for the use of disabled people (e.g. wheelchair-accessible unisex toilets, changing rooms or shower rooms) should:

- be fitted with light action privacy bolts;   BR-AP-M (5.4d)
- be capable of being opened using a force
  no greater than 20 N;
- have an emergency release mechanism   BR-AP-M (5.4e)
  so that they are capable of being opened
  outwards (from the outside) in case of
  emergency.

Door-opening furniture should:

- be easy to operate by people with limited   BR-AP-M (5.4c)
  manual dexterity;
- be easy to operate with one hand using a
  closed fist (e.g. a lever handle);
- contrast visually with the surface of the door.

Doors when open should not obstruct emergency escape routes.

Outlets, controls and switches

All controls and switches should be easy to operate, visible and free from obstruction and: • should be located between 750 mm and 1200 mm above the floor; • should not require the simultaneous use of both hands (unless necessary for safety reasons) to operate.	BR-AP-M (5.4i)
Light switches should: • have large push pads; • align horizontally with door handles; • be within 900 to 1100 mm from the entrance door opening.	BR-AP-M (5.4i)
Switched socket outlets should indicate whether they are 'ON'.	BR-AP-M (5.4i)
Mains and circuit isolator switches should clearly indicate whether they are 'ON' or 'OFF'.	BR-AP-M (5.4i)
Individual switches on panels and on multiple socket outlets should be well separated.	BR-AP-M (5.4i)
Controls that need close vision (e.g. thermostats) should be located between 1200 mm and 1400 mm above the floor.	BR-AP-M (5.4i)
Emergency alarm pull cords should: • be coloured red; • be located as close to a wall as possible; • have two red 50 mm diameter bangles.	BR-AP-M (5.4i)
Front plates should contrast visually with their backgrounds.	BR-AP-M (5.4i)
Heat emitters should either be screened or have their exposed surfaces kept at a temperature below 43°C.	BR-AP-M (5.4j)

Where possible, light switches with large push pads should be used in preference to pull cords. The colours red and green should not be used in combination as indicators of 'ON' and 'OFF' for switches and controls.	BR-AP-M (5.3)

Water closets

In large building complexes (e.g. retail parks and large sports centres) there should be one wheelchair-accessible unisex toilet capable of including an adult changing table.	BR-AP-M (5.17)
The dimensions of the self-contained compartment should allow sufficient space for a helper.	BR-AP-M (5.16)
A combined facility should be divided into distinct 'wet' and 'dry' areas.	BR-AP-M (5.16)

Toilets in separate-sex washrooms

There should be at least the same number of WCs (for women) as urinals (for men).	BR-AP-M (5.13)
Ambulant disabled people should have the opportunity to use a WC compartment within any separate-sex toilet washroom.	BR-AP-M (5.11)
A wheelchair-accessible compartment (where provided) shall have the same layout and fittings as the unisex toilet.	BR-AP-M (5.14f)
Where a separate-sex toilet washroom can be accessed by wheelchair users, it should be possible for them to use both a urinal (where appropriate) and a washbasin at a lower height than is provided for other users.	BR-AP-M (5.13)
Consideration should be given to providing a low-level urinal for children in male washrooms.	BR-AP-M (5.13)
Separate-sex toilet washrooms above a certain size should include an enlarged WC cubicle for use by people who need extra space, e.g. parents with children and babies, people carrying luggage and also ambulant disabled people.	BR-AP-M (5.12)
The minimum dimensions of compartments for ambulant disabled people should comply with Figure 12.8.	BR-AP-M (5.14b)

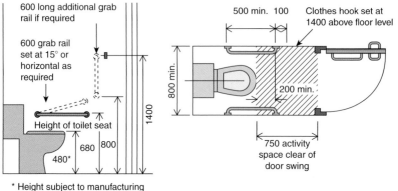

* Height subject to manufacturing
  tolerance of WC pan

Figure 12.8 Example of a WC cubicle for an ambulant disabled person

Unisex toilets

The approach to a unisex toilet should be separate to other sanitary accommodation.	BR-AP-M (5.9)
Wheelchair users should:	
• not have to travel more than 40 m on the same floor to reach a unisex toilet;	BR-AP-M (5.9 and 5.10h)
• not have to travel more than the combined horizontal distance where the unisex toilet accommodation is on another floor of the building (accessible by passenger lift);	BR-AP-M (5.9 and 5.10h)
• be able to approach, transfer to, and use the sanitary facilities provided within a building.	BR-AP-M (5.8)
Wheelchair-accessible unisex toilets should:	
• not be used for baby changing;	BR-AP-M (5.5)
• be located as close as possible to the entrance and/or waiting area of the building;	BR-AP-M (5.10a)
• not be located in a way that compromises the privacy of users;	BR-AP-M (5.10b)
• be located in a similar position on each floor of a multi-storey building;	BR-AP-M (5.10c)
• allow for right- and left-hand transfer on alternate floors;	BR-AP-M (5.10c and d)
• be located on accessible routes that are direct and obstruction free;	BR-AP-M (5.10f)
• always be provided in addition to any wheelchair-accessible accommodation in separate-sex toilet washrooms;	BR-AP-M (5.5)

The minimum overall dimensions of, and arrangement of fittings within, a wheelchair-accessible unisex toilet should comply with Figure 12.9.

BR-AP-M (5.10i)

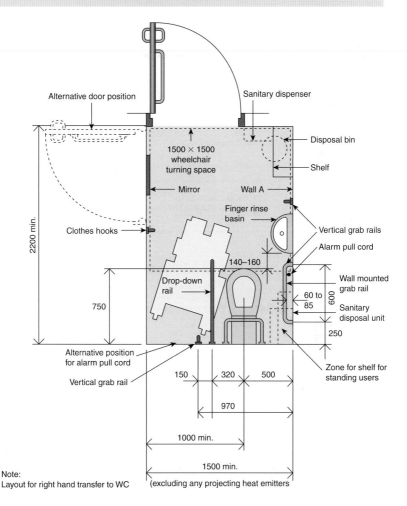

Alternative door position

Sanitary dispenser

Disposal bin

1500 × 1500 wheelchair turning space

Shelf

Mirror

Wall A

Finger rinse basin

Clothes hooks

Vertical grab rails

Alarm pull cord

2200 min.

140–160

Drop-down rail

Wall mounted grab rail

750

60 to 85

600

Sanitary disposal unit

250

Alternative position for alarm pull cord

150   320   500

Zone for shelf for standing users

Vertical grab rail

970

1000 min.

1500 min.

Note:
Layout for right hand transfer to WC   (excluding any projecting heat emitters)

Figure 12.9 Example of a unisex wheelchair-accessible toilet with a corner WC

Heights and arrangements

The heights and arrangement of fittings in a wheelchair-accessible unisex toilet should comply with Figure 12.10.

BR-AP-M (5.10)

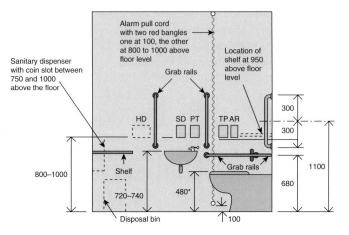

* Height subject to manufacturing tolerance of WC pan

HD: Possible position for automatic hand dryer
SD: Soap dispenser
PT: Paper towel dispenser
AR: Alarm reset button
TP: Toilet paper dispenser

* Height of drop-down rails to be the same as the other horizontal grab rails

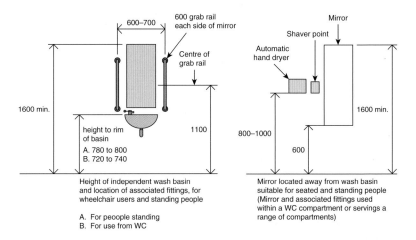

Height of independent wash basin
and location of associated fittings, for
wheelchair users and standing people

A. For peoople standing
B. For use from WC

Mirror located away from wash basin
suitable for seated and standing people
(Mirror and associated fittings used
within a WC compartment or servings a
range of compartments)

Figure 12.10 Typical heights of arrangement of fittings for a unisex,
wheelchair-accessible toilet

The space provided for manoeuvring should enable   BR-AP-M (5.8)
wheelchair users to adopt various transfer techniques
that allow independent or assisted use.
The transfer space alongside the WC should be kept   BR-AP-M (5.8)
clear to the back wall.

The relationship of the WC to the finger rinse basin and other accessories should allow a person to wash and dry hands while seated on the WC.   BR-AP-M (5.8)

Heat emitters (if located) should not restrict: BR-AP-M (5.10p)

• the minimum clear wheelchair manoeuvring space;
• the space beside the WC used for transfer from the wheelchair to the WC.

Doors

Doors to WC cubicles and wheelchair-accessible unisex toilets should:

• (ideally) open outwards;   BR-AP-M (5.3)
• be operable by people with limited strength or manual dexterity;
• be capable of being opened if a person has collapsed against them while inside the cubicle;
• be fitted with a horizontal closing bar fixed to the inside face.   BR-AP-M (5.10g)

Support rails

The rail on the open side can be a drop-down rail, but on the wall side, it can be either a wall-mounted grab rail or, alternatively, a second drop-down rail in addition to the wall-mounted grab rail.   BR-AP-M (5.8)

If the horizontal support rail (on the wall adjacent to the WC) is set at the minimum spacing from the wall, an additional drop-down rail should be provided on the wall side, 320 mm from the centre line of the WC.   BR-AP-M (5.10j)

If the horizontal support rail (on the wall adjacent to the WC) is set so that its centre line is 400 mm from the centre line of the WC, there is no additional drop-down rail.   BR-AP-M (5.10k)

Emergency assistance

An emergency assistance alarm system should be
provided that:

- has an outside emergency assistance call signal   BR-AP-M
  that can be easily seen and heard by those able   (5.10n)
  to give assistance;
- has visual and audible indicators to confirm that   BR-AP-M
  an emergency call has been received;   (5.10m)
- has a reset control reachable from a wheelchair,
  a WC, or from a shower/changing seat;
- has a signal that is distinguishable visually and
  audibly from the fire alarms provided.

Emergency assistance pull cords should:

- be easily identifiable;   BR-AP-M (5.10o)
- be reachable from the WC and from the floor
  close to the WC;
- be coloured red;
- be located as close to a wall as possible;
- have two red 50 mm diameter bangles.

Alarms

Fire alarms should emit an audio and visual signal to   BR-AP-M (5.4 g)
warn occupants with hearing or visual impairments.
Emergency assistance alarm systems should have: BR-AP-M (5.4 h)

- visual and audible indicators to confirm that
  an emergency call has been received;
- a reset control reachable from a wheelchair, a
  WC, or from a shower/changing seat;
- a signal that is distinguishable visually and
  audibly from the fire alarm.

WC pans

WC pans should: BR-AP-M (5.14e)

- conform to BS 5503-3 or BS 5504-4;
- accommodate the use of a variable height   BR-AP-M (5.10q)
  toilet seat riser;

- have a flushing mechanism positioned on the   BR-AP-M (5.10r)
  open or transfer side of the space, irrespective
  of handing.

Note: See BS 8300 for more detailed guidance on the various techniques used to transfer from a wheelchair to a WC, as well as appropriate sanitary and other fittings.

12.2.6 Provision of toilet accommodation

Where sanitary facilities are provided in a building,   BR-AP-M (5.7b)
at least one wheelchair-accessible unisex toilet
should be available.
If there is only space for one toilet in a building, then   BR-AP-M (5.7a)
it should be a wheelchair-accessible unisex type.
In separate-sex toilet accommodation, at least   BR-AP-M (5.7c)
one WC cubicle should be provided for ambulant
disabled people.
In separate-sex toilet accommodation having four or   BR-AP-M (5.7d)
more WC cubicles, at least one should be an enlarged
cubicle for use by people who need extra space.
Wheelchair-accessible unisex toilets should always   BR-AP-M (5.5)
be provided in addition to any wheelchair-accessible
accommodation in separate-sex toilet washrooms.
If there is only space for one toilet in a building:

- it should be a wheelchair-accessible unisex type;   BR-AP-M (5.7a)
- its width should be increased from 1.5 m to   BR-AP-M (5.7e)
  2 m;
- it should include a standing height washbasin,   BR-AP-M (5.7e)
  in addition to the finger rinse basin associated
  with the WC.

For specific guidance on the provision of sanitary accommodation in sports buildings, refer to the Disability Discrimination Act (DDA) 2005.

WC provision in the entrance storey of the dwelling

Whenever possible, a WC should be provided in the entrance storey of the dwelling so that there is no need to negotiate a stair to reach it from the habitable rooms in that storey.

If there is a bathroom in the principal storey, then a WC may be collocated with it.

The door to the WC compartment should:	BR-AP-M (10.3b)

- open outwards;
- be positioned so as to allow wheelchair users access to it;
- have a clear opening width in accordance with Table 12.8.

Table 12.8 Minimum widths of corridors and passageways for a range of doorway widths

Doorway clear opening width (mm)	Corridor/passageway width (mm)
750 or wider	900 (when approached head on)
750	1200 (when not approached head on)
775	1050 (when not approached head on)
800	900 (when not approached head on)

The WC compartment should:	BR-AP-M (10.3c)

- provide a clear space for wheelchair users to access the WC;
- position the washbasin so that it does not impede access.

Note: For further information see the Disability Rights Commission's website at http://www.drc-gb.org

Ambulant disabled people

Compartments used for ambulant disabled people should:	BR-AP-M
	(5.14d)

- be 1200 mm wide;
- include a horizontal grab bar adjacent to the WC;
- include a vertical grab bar on the rear wall;
- include space for a shelf and fold-down changing table.

A wheelchair-accessible washroom (where provided) shall have:	BR-AP-M
	(5.14g)

- at least one washbasin with its rim set at 720 to 740 mm above the floor;

- for men, at least one urinal with its rim set at
  380 mm above the floor, with two 600 mm long
  vertical grab bars with their centre lines at
  1100 mm above the floor, positioned either
  side of the urinal.

The compartment should:	BR-AP-M (5.11)

- be fitted with support rails;
- include a minimum activity space to accommodate
  people who use crutches, or otherwise have
  impaired leg movements.

Doors

Doors to compartments for ambulant disabled people should:	BR-AP-M (5.14d)

- preferably open outwards;
- be fitted with a horizontal closing bar fixed to the
  inside face.

The swing of any inward opening doors to standard WC compartments should enable a 450 mm diameter manoeuvring space to be maintained between the swing of the door, the WC pan and the side wall of the compartment.	BR-AP-M (5.14a)

Bathrooms and shower facilities

In a building where baths or showers are provided, accessible sanitary accommodation for disabled people should be provided at a ratio of one in 20 or part thereof, for each type of sanitary facility provided.	SBR-ND-3.12.10
An accessible bathroom intended for a disabled person should include a transfer space of at least 400 mm across the full width of the head of the bath.	SBR-ND-3.12.10
An accessible shower room intended for a disabled person should have:	SBR-ND-3.12.10

- a dished floor of a gradient of not more
  than 1:50 discharging into a floor drain, or a
  proprietary level access shower with a drainage
  area of not less than 1.2 m by 1.2 m; and
- a folding shower seat positioned 500 mm from
  a flanking wall and securely fixed, with a
  seat height that permits transfer to and from a
  wheelchair positioned outside the showering area.

Wheelchair-accessible bathrooms

Wheelchair users and ambulant disabled people (in hotels, motels, relatives'
accommodation in hospitals, student accommodation and sports facilities)
should be able to wash or bathe either independently or with assistance from
others.

The minimum overall dimensions of (and the arrangement of fittings within) a bathroom for individual use incorporating a corner WC should comply with Figure 12.11.	BR-AP-M (5.21a)

A choice of layouts suitable for left-hand and right-hand transfer should be provided when more than one bathroom is available.	BR-AP-M (5.21b)
The floor of a bathroom should be slip resistant when dry or when wet.	BR-AP-M (5.21c)
The bath should be provided with a transfer seat that is 400 mm deep and equal to the width of the bath.	BR-AP-M (5.21d)
Outward opening doors, fitted with a horizontal closing bar fixed to the inside face, should be provided.	BR-AP-M (5.21e)
An emergency assistance pull cord should be provided which should:	BR-AP-M (5.21f)

- be easily identifiable and reachable from the
  wall-mounted tip-up seat (or from the floor);
- be located as close to a wall as possible;
- be coloured red;
- have two red 50 mm diameter bangles.

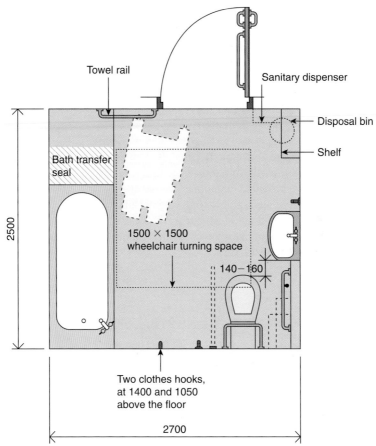

Towel rail

Sanitary dispenser

Disposal bin

Shelf

Bath transfer seal

2500

1500 × 1500 wheelchair turning space

140 – 160

Two clothes hooks, at 1400 and 1050 above the floor

2700

Note: Layout shown for right-hand transfer to bath and WC.

Figure 12.11 Example of a bathroom containing a WC

Notes:

- More detailed guidance on appropriate sanitary and other fittings, including facilities for the use of mobile and fixed hoists, is given in BS 8300.
- Guidance on the slip resistance of floor surfaces is given in Annex C of BS 8300.

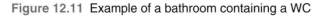

Shower facilities

A choice of shower layout together with correctly located shower controls and fittings will enable disabled people to make use of the facilities independently, or be assisted by others where necessary.

A shower curtain (enclosing the seat and rails and which can be operated from the shower seat) should be provided.	BR-AP-M (5.18m)
A shelf (that can be reached from the shower seat and/or from the wheelchair) should be provided for toiletries.	BR-AP-M (5.18n)
The floor of the shower and shower area should be slip resistant and self-draining.	BR-AP-M (5.18o)
The shower controls should be positioned between 750 and 1000 mm above the floor in all communal area wheelchair-accessible shower facilities.	BR-AP-M (5.18q)
If showers are provided in commercial developments for the benefit of staff, at least one wheelchair-accessible shower should be made available.	BR-AP-M (5.18l)
Individual self-contained shower facilities should comply with Figure 12.12.	BR-AP-M (5.18k)

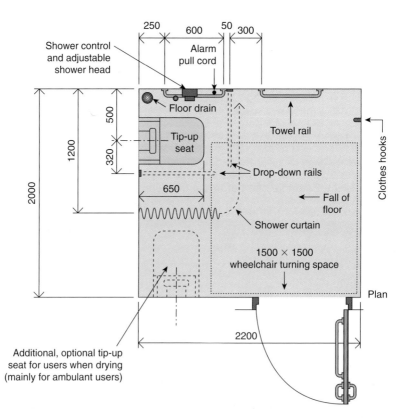

Figure 12.12 Example of a self-contained shower room for individual use

Shower facilities incorporating a WC

A choice of left-hand and right-hand transfer BR-AP-M (5.18s)
layouts should be available when more than one
shower area includes a corner WC.

The minimum overall dimensions of (and the BR-AP-M (5.18r)
arrangement of fittings within) an individual self-
contained shower area incorporating a corner
WC, e.g. in a sports building, should comply with
Figure 12.13.

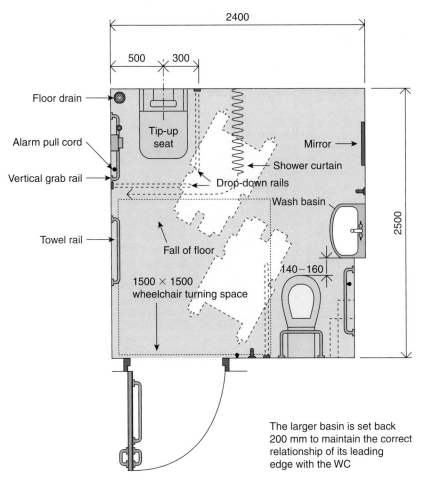

Note: Layout shown for right-hand transfer to shower seat and WC

Figure 12.13 Example of a shower room incorporating a corner WC for individual use

Note: More detailed guidance on appropriate sanitary and other fittings is given in BS 8300.

Changing facilities

Changing facilities intended for disabled people should be based on the following recommendations.

The floor of a changing area should be level and slip resistant when dry or when wet, particularly where it is associated with shower facilities.	BR-AP-M (5.18i)
There should be a manoeuvring space 1500 mm deep in front of lockers in self-contained and/or communal changing areas.	BR-AP-M (5.18j)
The minimum overall dimensions of (and the arrangement of equipment and controls within) individual self-contained changing facilities should comply with Figure 12.14.	BR-AP-M (5.18h)

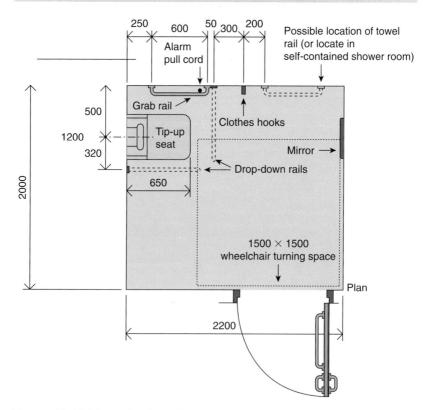

Figure 12.14 Example of a self-contained changing room for individual use

A choice of layouts suitable for left-hand and right-hand transfer should be provided when more than one individual changing compartment or shower compartment is available.

BR-AP-M (5.18a)

Wall-mounted drop-down support rails and wall-mounted slip-resistant tip-up seats (not spring-loaded) should be provided.

BR-AP-M (5.18b)

Subdivisions (with the same configuration of space and equipment as for self-contained facilities) should be provided for communal shower facilities and changing facilities.

BR-AP-M (5.18c)

In addition to communal separate-sex facilities, individual self-contained shower and changing facilities should be available in sports amenities.

BR-AP-M (5.18d)

An emergency assistance alarm system should be provided and should have:

BR-AP-M (5.18f)

- visual and audible indicators to confirm that an emergency call has been received;
- a reset control reachable from a wheelchair, a WC, or from a shower/changing seat;
- a signal that is distinguishable visually and audibly from the fire alarm.

An emergency assistance pull cord should be provided which should:

BR-AP-M (5.18e)

- be easily identifiable and reachable from the wall-mounted tip-up seat (or from the floor);
- be located as close to a wall as possible;
- be coloured red;
- have two red 50 mm diameter bangles.

Facilities for limb storage should be included for the benefit of amputees.

BR-AP-M (5.18)

13

Hot Water Storage Systems

13.1 Requirements

13.1.1 Hot water storage

A hot water storage system that has a hot water vessel which does not incorporate a vent pipe to the atmosphere shall be installed by a person competent to do so and there shall be precautions:
- to prevent the temperature of stored water at any time exceeding 100°C; and
- to ensure that the hot water discharged from safety devices is safely conveyed to where it is visible but will not cause danger to persons in or about the building.

Note: This requirement does not apply to:
- a hot water storage system that has a storage vessel with a capacity of 15 litres or less;
- a system providing space heating only;
- a system that heats or stores water for the purposes only of an industrial process.

13.1.2 Heating and hot water system(s)

If a heating or hot water system is being provided or extended, then the installed appliance should:

- not be less than that recommended for its type in the Domestic Heating Compliance Guide;
- have an efficiency which is not worse than 2 per cent lower than that of the appliance being replaced – if the appliance is the primary heating service;
- be provided with controls that meet the minimum control requirements of the Domestic Heating Compliance Guide for the particular type of appliance and heat distribution system;
- be commissioned so that at completion, the system(s) and their controls are left in working order and can operate efficiently for the purposes of the conservation of fuel and power.

The person carrying out the work shall provide the local authority with a notice (signed by a suitably qualified person) confirming that all fixed building services have been properly commissioned in accordance with the *Domestic Heating Compliance Guide*.

13.1.3 Insulation of pipes, ducts and vessels

Hot and chilled water pipework, storage vessels, refrigerant pipework and ventilation ductwork should be insulated so as to conserve energy and to maintain the temperature of the heating or cooling service.

13.1.4 Building services (non-domestic buildings)

For energy-saving purposes:

- all heating, ventilation and air-conditioning systems should be provided with controls to enable them to achieve reasonable standards of energy efficiency;
- heating and hot water service systems (cooling plant and/or air-handling plants) should have an efficiency not less than that recommended for its type in the Non-Domestic Heating, Cooling and Ventilation Compliance Guide.

13.1.5 Health and safety

Every building must be designed and constructed in such a way that protection is provided for people in, and around, the building from the danger of severe burns or scalds from the discharge of steam or hot water.

The hot water system shall:

- be installed by a competent person;
- not exceed 100°C;
- discharge safely;
- not cause danger to persons in or about the building.

13.1.6 Energy efficiency

Every building must be designed and constructed in such a way that:

- the heating and hot water service systems installed are energy efficient and are capable of being controlled to achieve optimum energy efficiency;
- temperature loss from heated pipes, ducts and vessels, and temperature gain to cooled pipes and ducts, is resisted.

13.1.7 Commissioning of heating and hot water systems

When heating and hot water systems are commissioned:

- the systems and their controls shall be left in their intended working order and should operate efficiently for the purposes of the conservation of fuel and power;
- the person carrying out the work shall provide the local authority with a notice (signed by a suitably qualified person) confirming that all fixed building services have been properly commissioned in accordance with the Domestic Heating Compliance Guide.

13.1.8 Information

The occupiers of a building must be provided with written information by the owner on the operation and maintenance of the building services and energy supply systems.

13.2 Meeting the requirements

In domestic buildings it is normal that systems of up to 500 litres storage capacity and a power input not exceeding 45 kW are used.

13.2.1 Heating and hot water systems

Every unvented water heater, not being an instantaneous water heater with a capacity not greater than 15 litres, and every secondary coil contained in a primary system shall:	SI 1148
• be fitted with a vent pipe, a temperature control device, a temperature relief device and a combined temperature pressure and relief valve; or	17(1)(a)
• be capable of accommodating expansion within the secondary hot water system.	17(1)(b)
An expansion valve shall be fitted with provision to ensure that water is discharged in a correct manner in the event of a malfunction of the expansion vessel or system.	SI 1148-17(2)

Appropriate vent pipes, temperature control devices and combined temperature pressure and relief valves shall be provided to prevent the temperature of the water within a secondary hot water system from exceeding 100°C.	SI 1148-18
Discharges from temperature relief valves, combined temperature pressure and relief valves and expansion valves shall be made in a safe and conspicuous manner.	SI 1148-19
No vent pipe from a primary circuit shall terminate over a storage cistern containing wholesome water for domestic supply or for supplying water to a secondary system.	SI 1148-20(1)
No vent pipe from a secondary circuit shall terminate over any combined feed and expansion cistern connection to a primary circuit.	SI 1148-20(2)
Every expansion cistern or expansion vessel, and every cold water combined feed and expansion cistern connected to a primary circuit, shall be such as to accommodate any expansion water from that circuit during normal operation.	SI 1148-21
Every expansion valve, temperature relief valve or combined temperature and pressure relief valve connected to any fitting or appliance shall close automatically after a discharge of water.	SI 1148-22(1)
Every expansion valve shall:	SI 1148
• be fitted on the supply pipe close to the hot water vessel and without any intervening valves; and	22(2)(a)
• only discharge water when subjected to a water pressure of not less than 0.5 bar (50 kPa) above the pressure to which the hot water vessel is, or is likely to be, subjected in normal operation.	22(2)(b)
A temperature relief valve or combined temperature and pressure relief valve shall be provided on every unvented hot water storage vessel with a capacity greater than 15 litres.	SI 1148-23(1)
The valve shall:	SI 1148
• be located directly on the vessel in an appropriate location, and have a sufficient discharge capacity, to ensure that the temperature of the stored water does not exceed 100°C; and	23(2)(a)

- only discharge water at below its operating   23(2)(b)
  temperature when subjected to a pressure of not
  less than 0.5 bar (50 kPa) in excess of the greater
  of the following:
  ○ the maximum working pressure in the vessel
    in which it is fitted, or
  ○ the operating pressure of the expansion valve.

Note: 'Unvented hot water storage vessel' means a hot water storage vessel that does not have a vent pipe to the atmosphere.

No supply pipe or secondary circuit shall be   SI 1148-24
permanently connected to a closed circuit for filling
a heating system unless it incorporates a backflow
prevention device in accordance with a specification
approved by the Regulator/Scottish Ministers (as
applicable) for the purposes of this Schedule.

Backflow prevention

Every water system shall contain an adequate device   SI 1148-15(1)
or devices for preventing backflow of fluid from any
appliance, fitting or process from occurring unless it is:

- a water heater where the expanded water is   15(2)(a)
  permitted to flow back into a supply pipe; or
- a vented water storage vessel supplied from a   15(2)(b)
  storage cistern;

where the temperature of the water in the supply pipe
or the cistern does not exceed 25°C.
The device used to prevent backflow shall be   SI 1148-15(3)
appropriate to the highest applicable fluid category
to which the fitting is subject downstream before the
next such device.
Backflow prevention shall be provided on any supply   SI 1148
pipe or distributing pipe:

- where it is necessary to prevent backflow   15(4)(a)
  between separately occupied premises; or

- where the water undertaker has given notice for   15(4)(b)
  the purposes of this Schedule that such prevention
  is needed for the whole or part of any premises.

The device used to prevent backflow shall be   SI 1148-15(3)
appropriate to the highest applicable fluid category to
which the fitting is subject downstream before the next
such device.

Appliance efficiency

Table 13.1 provides details of the recommended minimum thermal efficiencies for domestic hot water systems.

 There is no minimum thermal efficiency specified for electric domestic hot water heaters.

Table 13.1 Domestic hot water systems

System type		Minimum thermal efficiency[a]
Direct-firing	Natural gas	73%
	LPG	74%
	Oil	75%
Indirect-firing	Natural gas	80%
	LPG	81%
	Oil	82%

[a]Based on gross calorific value.

13.2.2 Domestic hot water heating controls

Note: Although this guidance refers only to non-domestic buildings, hot water systems, to confuse matters, are generally referred to as domestic hot water (DHW) systems.

A DHW system should have controls that will switch   SBR-ND-6.3.7
off the heat when the water temperature required by
the occupants has been achieved and during periods
when there is no demand for hot water.
The controls shown in Tables 13.2 and 13.3 should be   SBR-ND-3.7
observed for all DHW systems.

Gas/oil-firing systems

Table 13.2 **Gas/oil-firing systems**

System	Controls
Direct	Automatic thermostat control to shut off the burner/primary heat supply when the desired temperature of the hot water has been reached
Indirect	Automatic thermostat control to shut off the burner/primary heat supply when the desired temperature of the hot water has been reached High limit thermostat to shut off primary flow if system temperature gets too high Time control

Table 13.3 **Electric domestic hot water systems**

Control system	Point of use	Local	Central	Instantaneous
Automatic thermostat control to interrupt the electrical supply when the desired storage temperature has been reached	Yes	Yes	Yes	No
High-limit thermostat (thermal cut-out) to interrupt the energy supply if the system temperature gets too high	Yes	Yes	Yes	No
Manual reset in the event of an over-temperature trip	Yes	Yes	Yes	No
A 7-day time-clock or building management system (BMS) interface should be provided to ensure bulk heating of water using off-peak electricity	No	Yes	Yes	No
High-limit thermostat (thermal cut-out) to interrupt the energy supply if the outlet temperature gets too high	No	No	No	Yes
Flow sensor that only allows electrical input should sufficient flow through the unit be achieved	No	No	No	Yes

Electric domestic hot water systems

A DHW system (other than a system with a solid   SBR-ND-6.3.7
fuel boiler) should have controls that will switch off
the heat when the water temperature required by the
occupants has been achieved and during periods when
there is no demand for hot water.

Independent time and temperature control of hot water circuits should be provided along with a boiler interlock (refer to table below; SBR-D-6.3.10) to ensure that the boiler and pump only operate when there is a demand for heat.	SBR-D-6.3.8
For hot water systems in large dwellings, more than one hot water circuit each with independent time and temperature control should be provided.	SBR-D-6.3.8
A hot water system (other than for combi boilers with storage capacity 15 litres or less) should have controls that will switch off the heat when the water temperature required by the occupants has been achieved and during periods when there is no demand for hot water.	SBR-D-6.3.8

Controls for dry space heating and hot water systems

For large dwellings with a floor area over $150\,m^2$:

• independent time and temperature control of multiple space heating zones is recommended;	SBR-D-6.3.9
• each zone (not exceeding $150\,m^2$) should have a room thermostat, and a single multi-channel programmer or multiple heating zone programmers.	SBR-D-6.3.9

Zone controls are not considered necessary for single-apartment dwellings.

Controls for combined warm air and hot water systems

The following controls should be provided:	SBR-D-6.3.10
• independent time control of both the heating and hot water circuits (achieved by means of a cylinder thermostat and a timing device, wired such that when there is no demand for hot water both the pump and circulator are switched off);	
• pumped primary circulation to the hot water cylinder;	

- a hot water circulator interlock (achieved by means of a cylinder thermostat and a timing device, wired such that when there is no demand from the hot water both the pump and circulator are switched off); and
- time control by the use of either:
  - a full programmer with separate timing to each circuit;
  - two or more separate timers providing timing control to each circuit;
  - a programmable room thermostat(s) to the heating circuit(s); or
  - a time switch/programmer (two channels) and room thermostat.

Heat pumps for hot water systems

Heat pump unit controls should include: SBR-D-6.3.8

- control of water temperature for the distribution system;
- control of water pumps (integral or otherwise);
- defrost control of external airside heat exchanger (for air to water units);
- control of outdoor fan operation (for air to water units);
- protection for water flow failure;
- protection for high water temperature;
- protection for high refrigerant pressure; and
- protection for external air flow failure (on air to water units).

Controls which are not integral to the unit should include:

- room thermostat to regulate the space temperature and interlocked with the heat pump unit operation; and
- timer to optimize operation of the heat pump.

13.2.3 Insulation of pipes, ducts and vessels

Insulation of hot water pipes

Hot water pipes serving a space heating system should be thermally insulated against heat loss unless the use of such pipes or ducts always contributes to the heating demands of the room or space.	SBR-D-6.4.1, SBR-ND-6.4.1
Pipes that are used to supply hot water to appliances within a building should be insulated against heat loss.	SBR-D-6.4.1, SBR-ND-6.4.1

For buildings other than dwellings:

• insulation should not be less than those shown in the *Non-Domestic Heating, Cooling and Ventilation Compliance Guide;*	BR-AP-L1A 49 and L1B 39
• hot and chilled water pipework, storage vessels, refrigerant pipework and ventilation ductwork should be insulated so as to conserve energy and to maintain the temperature of the heating or cooling service.	BR-AP-2B 52

It is not intended that the following guidance is applied to storage systems with a capacity of less than 15 litres, to systems used solely for space heating or to any system used for an industrial or a commercial process.

Every building must be designed and constructed in such a way that protection is provided for people in, and around, the building from the danger of severe burns or scalds from the discharge of steam or hot water.	SBR-D-4.9.0
Safety devices installed to protect from hazards such as scalding or the risk of explosion of unvented systems should be maintained to ensure correct operation.	SBR-D-4.9.0

Insulation of vessels

Where an unvented hot water system is installed, additional insulation should be considered to reduce the heat loss that can occur from the safety fittings and pipework.	SBR-D-6.4.2, SBR-ND-6.4.2
A hot water storage vessel should be insulated against heat loss.	SBR-D-6.4.2, SBR-ND-6.4.2

13.2.4 Unvented hot water systems

Small unvented hot water storage systems: specification

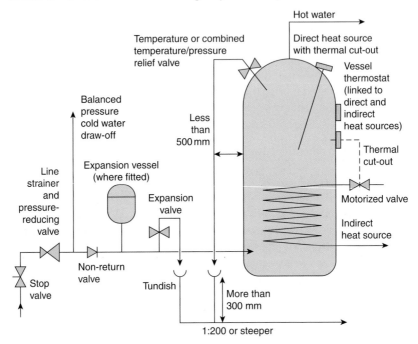

Figure 13.1 Unvented hot water storage system: indirect example

An unvented hot water storage system (Figure 13.1) should be:

• designed and installed to prevent the temperature of the stored water at any time exceeding 100°C and to provide protection from malfunctions of the system; and	SBR-D-4.9.2, SBR-ND-4.9.2
• in accordance with the recommendations of BS 7206:1990.	

A unit or package should have fitted (or supplied for fitting by the installer):	SBR-D-4.9.2, SBR-ND-4.9.2

- • a check valve to prevent backflow; and
- • a pressure control valve to suit the operating pressure of the system; and
- • an expansion valve to relieve excess pressure; and
- • an external expansion vessel or other means of accommodating expanded heated water.

Additional to any thermostatic control that is fitted to maintain the temperature of the stored water at around 60°C, a unit or package should have a minimum of two temperature-activated devices operating in sequence comprising:

SBR-D-4.9.2, SBR-ND-4.9.2

- a non-self-resetting thermal cut-out; and
- a temperature relief valve.

A temperature-operated, non-self-resetting, energy cut-out complying with BS 3955:1986 should be fitted to the vessel.

SBR-D-4.9.2, SBR-ND-4.9.2

In the event of thermostat failure, heating to the water in the vessel should stop before the temperature rises to the critical level required for operation of the temperature relief valve.

In indirectly heated vessels, the non-self-resetting thermal cut-out should operate a motorized valve, or other similar device, to shut off the flow from the heat source.

SBR-D-4.9.2, SBR-ND-4.9.2

On directly heated vessels or where an indirectly heated vessel has an alternative direct method of water heating fitted, a non-self-resetting thermal cut-out device should be provided for each direct source.

SBR-D-4.9.2, SBR-ND-4.9.2

The temperature relief valve should be located directly on the storage vessel.

SBR-D-4.9.2, SBR-ND-4.9.2

The relief valve should have a discharge capacity rating at least equal to the rate of energy (power in kilowatts) input to the heat source.

SBR-D-4.9.2, SBR-ND-4.9.2

Large unvented hot water storage systems: specification

An unvented hot water storage system should be designed and installed to prevent the temperature of the stored water at any time exceeding 100°C and to provide protection from malfunctions of the system.

Where the system has a power input of less than 45 kW, safety devices (such as thermal cut-outs and discharge pipework) should be provided.

SBR-D 4.9.3

> Where the system has a power input greater than 45 kW, safety devices should also include temperature or combined temperature/pressure relief valves capable of a combined discharge rating at least equal to the power input of the system.
>
> SBR-ND-4.9.3

Installation of unvented hot water storage systems

The installation of an unvented hot water storage system should be carried out by a person with appropriate training and practical experience, including current membership of a registration scheme operated by a recognized professional body such as the Scottish and Northern Ireland Plumbing Employers' Federation (SNIPEF) or the Construction Industry Training Board (CITB).

Concerning the installation of an unvented hot water storage system:

> - the installer should be a competent person and, on completion, the labelling of the installation should identify the installer;
> - the installed system should be meet the recommendations of BS 7206:1990 or be the subject of an approval by a notified body;
> - certification of the unit or package should be recorded by permanent marking and a warning label which should be visible after installation. A comprehensive installation/user manual should be supplied;
> - the tundish and discharge pipework should be correctly located and fitted by the installer and the final discharge point should be visible and safely positioned where there is no risk from hot water discharge;
> - the operation of the system under discharge conditions should be tested to ensure provision is adequate.
>
> SBR-D-4.9.1,
> SBR-ND-4.9.1

13.2.5 Discharge from unvented hot water storage systems

The removal of discharges of water from the system can be considered in three parts.

Relief valve to tundish

Each relief valve should discharge into a metal pipe not less than the nominal outlet size of the valve. The discharge pipe should have an air-break, such as a tundish, not more than 500 mm from the vessel relief valve and located in an easily visible location within the same enclosure.	SBR-D-4.9.3, SBR-ND-4.9.4 SBR-D-4.9.3, SBR-ND-4.9.4

Discharge pipes from more than one relief valve may be taken through the same tundish.

Pipework should be installed so that any discharge will be directed away from electrical components should the discharge outlet become blocked.	SBR-D-4.9.3, SBR-D-4.9.4

Tundish to final discharge point

The discharge pipe from the tundish to final discharge point should be of a material, usually copper, capable of withstanding water temperatures of up to 95°C and be at least one pipe size larger than the outlet pipe to the relief valve.	SBR-D-4.9.3, SBR-ND-4.9.4
A vertical section of pipe, at least 300 mm long, should be provided beneath the tundish before any bends to the discharge pipe; thereafter the pipe should be appropriately supported to maintain a continuous fall of at least 1 in 200 to the discharge point.	SBR-D-4.9.3, SBR-ND-4.9.4
The pipework should have a resistance to the flow of water no greater than that of a straight pipe 9 m long unless the pipe bore is increased accordingly. Guidance on sizing of pipework from the tundish to the final discharge point is shown in Table 13.4.	SBR-D-4.9.3, SBR-ND-4.9.4

Table 13.4 **Size of discharge pipework**

Valve outlet size	Minimum size of discharge pipe to tundish	Minimum size of discharge pipe from tundish	Maximum resistance allowed[a]	Equivalent resistance created by the addition of each elbow or bend
G1/2	15 mm	22 mm	Up to 9 m	0.8 m
		28 mm	Up to 18 m	1.0 m
		35 mm	Up to 27 m	1.4 m
G3/4	22 mm	28 mm	Up to 9 m	1.0 m
		35 mm	Up to 18 m	1.4 m
		42 mm	Up to 27 m	1.7 m
G1	28 mm	35 mm	Up to 9 m	1.4 m
		42 mm	Up to 18 m	1.7 m
		54 mm	Up to 27 m	2.3 m

[a]Expressed as a length of straight pipe, i.e. no elbows or bends.

Discharge pipe termination

The pipe termination should be in a visible location and installed so that discharge will not endanger anyone inside or outside the building.

SBR-D-4.9.3, SBR-ND-4.9.4

Ideally, the final discharge point should be above the water seal to an external gully and below a fixed grating. Other methods for terminating the final discharge point would include:

SBR-D-4.9.3, SBR-ND-4.9.4

- up to 100 mm above external surfaces such as car parks, grassed areas or hard standings; a wire cage or similar guard should be provided to both prevent contact with discharge and protect the outlet from damage, whilst maintaining visibility;
- at high level into a hopper and downpipe of a material, such as cast iron, appropriate for a hot water discharge with the end of the discharge pipe clearly visible;
- onto a flat roof or pitched roof clad in a material capable of withstanding high-temperature discharges of water, such as slate/clay/concrete tiles or metal sheet, with the discharge point a minimum of 3 m from any plastic guttering system that would collect such discharges.

Note: Discharge at high level may be possible if the discharge outlet is terminated in such a way as to direct the flow of water against the external face of a wall. However, evidence of the minimum height of the outlet above any surface to which people have access and the distance needed to reduce the discharge to a non-scalding level should be established by test or otherwise.

13.2.6 Discharge of steam or hot water

In a domestic building, any vent or overflow pipe of a hot water system should be positioned so that any discharge will not endanger anyone inside or outside the building.	SBR-D-4.9.4
To prevent the development of *Legionella* or similar pathogens, hot water within a storage vessel should be stored at a temperature of not less than 60°C and distributed at a temperature of not less than 55°C.	SBR-D-4.9.5, SBR-ND-4.9.5

Hot water discharge from sanitary fittings

To prevent scalding, the temperature of hot water, at point of delivery to a bath or bidet, should not exceed 48°C.	SBR-D-4.9.5, SBR-ND-4.9.5
A device or system limiting water temperature should allow flexibility in setting of a delivery temperature, up to a maximum of 48°C, in a form that is not easily altered by building users.	SBR-D-4.9.5, SBR-ND-4.9.5

13.2.7 Solar water heating

Solar water heating has low or zero carbon dioxide emissions and low or no associated running costs, and is inherently energy efficient.

Solar roof panels should be regarded as forming part of the roof covering and as such should be able to resist ignition from an external source.	SBR-D-2.8, SBR-ND-2.8
Location and orientation for optimum energy efficiency and to avoid overshading should be considered.	SBR-D-6.3.6
All pipes of a solar water heating primary system should be insulated.	SBR-D-6.4.1, SBR-ND-6.4.1

Controls for solar water heating

Controls should be provided to:	SBR-D-6.3.11

- optimize the useful energy gain from the solar collectors into the system's dedicated storage vessel(s);
- minimize the accidental loss of stored energy by the solar hot water system;
- ensure that hot water produced by auxiliary heat sources is not used when adequate grade solar preheated water is available;
- guard against the adverse affects of excessive primary temperatures and pressures;
- limit the inlet temperature of any separate domestic hot water heating appliance (such as a combi boiler).

13.2.8 Solid fuel boilers

These should be thermostatically controlled to reduce the burning rate of the fuel, by varying the amount of combustion air to the fire.

For safety reasons, a suitable heat bleed (slumber circuit) from the system should be formed (e.g. a gravity-fed radiator without a thermostatic radiator valve or a hot water cylinder that is connected independently of any controls).	SBR-D-6.3.8
For hot water systems, unless the cylinder is forming the slumber circuit, a thermostatically controlled valve should be fitted.	SBR-D-6.3.8

13.2.9 Energy efficiency

Every building must be designed and constructed in such a way that the hot water systems installed are energy efficient and are capable of being controlled to achieve optimum energy efficiency.	SBR-D-6.3, SBR-ND-6.3

Dry central heating systems

Where a gas-fired circulator is incorporated in the warm-air unit to provide domestic hot water, it should be of a type that is able to deliver full- and part-load efficiency.	SBR-D-6.3.5

13.2.10 Work on existing buildings

Where alterations are being made to an existing heating/hot water system or a new or replacement heating/hot water system is being installed in an existing dwelling (or building consisting of dwellings), such alterations should not allow the heating system as a whole to be downgraded in terms of energy efficiency or compromised from a safety point of view.	SBR-D-6.3.12

Where a new or replacement boiler or hot water storage vessel is installed, or where existing systems are extended, new or existing pipes that are accessible or exposed as part of the work should be insulated as for new systems.	SBR-D-6.4.3, SBR-ND-6.4.3

When heating and hot water systems are commissioned:

• the performance of the building fabric and the heating and hot water systems should be no worse than the design limits;	BR-AP-L2A (Criterion 2) 9
• the heating and hot water system(s) should be commissioned so that at completion, the system(s) and their controls are left in working order and can operate efficiently for the purposes of the conservation of fuel and power.	BR-AP-L1A 69 and L1B 36

13.2.11 Written information

For a domestic building, written information   SBR-D-6.8.1
concerning the operation and maintenance of the hot
water system should be made available for the use of
the occupier.
For non-domestic buildings, a logbook containing   SBR-ND-6.8.1
information of energy system operation and
maintenance (such as building services plant and
controls) to ensure that the building user can optimize
the use of fuel, shall be provided.

14

Cold Water Storage Systems

14.1 Requirements

Every pipe supplying water connected to a storage cistern shall be fitted with an effective adjustable valve capable of shutting off the inflow of water at a suitable level below the overflowing level of the cistern.   SI 1148-16(1)

Every inlet to a storage cistern, combined feed and expansion cistern, WC flushing cistern or urinal flushing cistern shall be fitted with a servicing valve on the inlet pipe adjacent to the cistern.   SI 1148-16(2)

Every storage cistern, except one supplying water to the primary circuit of a heating system, shall be fitted with a servicing valve on the outlet pipe.   SI 1148-16(3)

Every storage cistern shall be fitted with: SI 1148

- *an overflow pipe, with a suitable means of warning of an impending overflow, which excludes insects;*   16(4)(a)
- *a cover positioned so as to exclude light and insects; and*   16(4)(b)
- *thermal insulation to minimize freezing or undue warming.*   16(4)(b)

Every storage cistern shall be so installed as to minimize the risk of contamination of stored water. The cistern shall be of an appropriate size, and the pipe connections to the cistern shall be so positioned as to allow free circulation and to prevent areas of stagnant water from developing.   SI 1148-16(5)

15

Heating Systems

In the design of all buildings, the energy efficiency of the heating plant is an important part of the package of measures that contributes to the overall carbon dioxide emissions from a building.

Ideally, the system should have sufficient zone, time and temperature controls to ensure that the heating system only provides the desired temperature when the building is occupied.

15.1 Requirements

Every building must be designed and constructed in such a way that:

- *the air quality inside the building is not a threat to the health of the occupants or the capability of the building to resist moisture, decay or infestation;* SBR-D-3.14, SBR-ND-3.14

- *the products of combustion are carried safely to the external air without harm to the health of any person through leakage, spillage or exhaust, nor permit the re-entry of dangerous gases from the combustion process of fuels into the building;* SBR-D-3.20, SBR-ND-3.20

- *the heating and hot water service systems installed are energy efficient and are capable of being controlled to achieve optimum energy efficiency;* SBR-D-6.3, SBR-ND-6.3

- *temperature loss from heated pipes, ducts and vessels, and temperature gain to cooled pipes and ducts, is resisted;* SBR-D-6.4, SBR-ND-6.4

- *an Energy Performance Certificate for the building is affixed to the building, indicating the approximate annual carbon dioxide emissions and energy usage of the building based on a standardized use of the building.* SBR-D-6.9, SBR-ND-6.9

> *The occupiers of a building must be provided with written*   SBR-D-6.8,
> *information by the owner on the operation and maintenance*   SBR-ND-6.8
> *of the building services and energy supply systems.*

15.2 Meeting the requirements

Heating, ventilation and thermal insulation should be considered as part of a total design that takes into account all heat gains and losses. Failure to do so can lead to inadequate internal conditions (e.g. condensation and mould and the inefficient use of energy due to overheating).

> Every building must be designed and constructed in   SBR-D-3.13
> such a way that it can be heated.
> Every dwelling should have some form of fixed   SBR-D-3.13.1
> heating system or alternative that is capable of
> maintaining a temperature of 21°C in at least one
> apartment and 18°C elsewhere, when the outside
> temperature is minus 1°C.

There is no need to maintain these temperatures in storage rooms with a floor area of not more than 4 m².

Alternative heating systems may involve a holistic design approach to the dwelling and can include the use of natural sources of available energy such as the sun, wind and the geothermal capacity of the earth.

> Where there are elderly or infirm occupants in a   SBR-D-3.13.2
> dwelling the capability of the heating system to
> maintain an apartment at a temperature higher than
> 21°C is a sensible precaution.

15.2.1 Heating appliances

Safe operation

The correct installation of heating appliances (and the design and installation of a flue) can reduce the risk from combustion appliances and their flues from:

- endangering the health and safety of persons in or around a building;
- compromising the structural stability of a building;
- causing damage by fire.

Domestic space and water heating facilities should be designed in accordance with Table 15.1.

Table 15.1 **Recommended designation for chimneys and flue pipes for use with oil-firing appliances with a flue gas temperature not more than 250°C**

Appliance type	Fuel oil	Designation
Boiler including combination boiler: pressure jet burner	Class C2	T250 N2 D 1 Oxx
Cooker: pressure jet burner	Class C2	T250 N2 D 1 Oxx
Cooker and room heater: vaporizing burner	Class C2	T250 N2 D 1 Oxx
Cooker and room heater: vaporizing burner	Class D	T250 N2 D 2 Oxx
Condensing pressure jet burner appliances	Class C2	T160 N2 W 1 Oxx
Cooker: vaporizing burner appliances	Class D	T160 N2 W 2 Oxx

Removal of products of combustion

Heating appliances fuelled by solid fuel, oil or gas all have the potential to cause carbon monoxide (CO) poisoning if they are poorly installed or commissioned, inadequately maintained or incorrectly used.

Insulation of pipes, ducts and vessels

Hot and chilled water pipework, storage vessels, refrigerant pipework and ventilation ductwork should be insulated so as to conserve energy and to maintain the temperature of the heating or cooling service.

Thermal insulation to heating pipes and ducts will improve energy efficiency by preventing:

- uncontrolled heat loss from such equipment;
- an uncontrolled rise in the temperature of the parts of the building where such equipment is situated.

Every building must be designed and constructed in such a way that temperature loss from heated pipes, ducts and vessels, and temperature gain to cooled pipes and ducts, is resisted.	SBR-D-6.4, SBR-ND-6.4

15.2.2 Heating systems

In the design of all buildings, the energy efficiency of the heating plant is an important part of the package of measures that contributes to the overall carbon dioxide emissions of the building.

Ideally, the system should have sufficient zone, time and temperature controls to ensure that the heating system only provides the desired temperature when the building is occupied.

Every building must be designed and constructed in such a way that the heating and hot water service systems installed are energy efficient and are capable of being controlled to achieve optimum energy efficiency.	SBR-D-6.3, SBR-ND-6.3
In non-domestic buildings:	SBR-ND-6.3.0

- a heating system boiler should be correctly sized to ensure energy efficiency;
- where future heating capacity is required, consideration should be given to providing additional space for extra plant; and
- pipework or ductwork should be configured to allow for the future loading.

15.2.3 Heating and hot water systems

If a heating or hot water system is being provided or extended then the installed appliance should:

- not be less than that recommended for its type in the Domestic Heating Compliance Guide;	BR-APs
- have an efficiency which is not worse than 2% lower than that of the appliance being replaced – if the appliance is the primary heating service;	
- be provided with controls that meet the minimum control requirements of the Domestic Heating Compliance Guide for the particular type of appliance and heat distribution system.	

Note: The person carrying out the work shall provide the local authority with a notice (signed by a suitably qualified person) confirming that all fixed building services have been properly commissioned in accordance with the *Domestic Heating Compliance Guide*.

Wet central heating efficiency

Gas and oil

Boilers and appliances installed in a dwelling or building consisting of dwellings should have minimum appliance efficiencies as shown in Table 15.2.	SBR-D-6.3.1

Table 15.2 **Gas and oil wet central heating efficiency**

Heating system	Efficiency (Gross calorific value)
Gas and oil central heating boilers (natural gas or LPG)	SEDBUK 86%, i.e. condensing boiler
Gas or oil (twin burner) range cooker central heating boilers (www.rangeefficiency.org.uk)	SEDBUK 75%
Gas-fired fixed independent space heating appliances used as primary space heating	58% gross
Oil-fired fixed independent space heating appliances used as primary space heating	60% gross

Each appliance should be capable of providing independent temperature control in areas with different heating needs. This could be independent or in conjunction with room thermostats or other appropriate temperature-sensing devices.

SBR-D-6.3.8

Solid fuel

The appliance efficiency should be at least that required for its category as designated by the Heating Equipment Testing Approval Scheme (HETAS) as given in Table 15.3.

SBR-D-6.3.2

Table 15.3 **Solid fuel wet central heating efficiency**

Category	Appliance type	Efficiency (gross calorific value)
D	Open fires with high-output boilers	63% (trapezium) 65% (rectangle)
F	Room heaters and stoves with boilers	67%
G	Cookers with boilers	50% (not more than 3.5 kW) 60% (3.5–7.5 kW)
J	Independent boilers (including pellet and log boilers)	65% (batch fed)
		70–75% (automatic anthracite)

Electric

For the most efficient use of electrical supplies it is recommended that an electric flow boiler is used to provide space heating alone, with the bulk of the hot water demand of the dwelling being supplied by a directly heated water heater utilizing 'off-peak' electricity tariffs.

Electric flow boilers should be constructed to meet   SBR-D-6.3.3
the requirements of the Low Voltage Directive and
Electromagnetic Compatibility Directive, preferably
shown by a third party electrical approval, e.g. British
Electrotechnical Approvals Board (BEAB) or similar.
Vented copper hot water storage vessels associated with   SBR-D-6.3.3
the system should meet BS 1566:2002 or BS 3198:1981.

Controls for wet space heating systems

Independent time and temperature control of heating   SBR-D-6.3.8
circuits should be provided along with a boiler
interlock (see Table 15.4) to ensure that the boiler and
pump only operate when there is a demand for heat.
Zone controls are not considered necessary for single-   SBR-D-6.3.8
apartment dwellings.
For large dwellings with a floor area over $150\,m^2$,   SBR-D-6.3.8
independent time and temperature control of multiple
space heating zones is recommended.
Each zone (not exceeding $150\,m^2$) should have a room   SBR-D-6.3.8
thermostat, and a single multi-channel programmer or
multiple heating zone programmers.

Table 15.4 **Controls for combination boilers, CPSU boilers and electric boilers**

Type of control	Means to achieve
Boiler control	Boiler interlock
	Automatic bypass valve
Time control	Time switch (7 day for space heating)
	Full programmer for electric
Room temperature control	TRVs (all radiators except in rooms with room thermostats or where 'heat bleed' required)
	Room thermostat(s)

CPSU: combined primary storage unit; TRV: thermostatic radiator valve.

An electric flow boiler should be fitted with a flow   SBR-D-6.3.8
temperature control and be capable of modulating the
power input to the primary water depending on space
heating conditions.

Controls for other boilers are summarized in Table 15.5.

Table 15.5 **Controls for other boilers**

Type of control	Means to achieve
Boiler control	Boiler interlock (for solid fuel as advised by manufacturer) Automatic bypass valve
Time control	Full programmer (7 day for space and hot water)
Room temperature control	As in Table 15.4
Cylinder control	Cylinder thermostat plus two-port valves or a three-port valve Separately controlled circuits to cylinder and radiators with pumped circulation
Pump control	Pump overrun timing device as required by manufacturer

Dry central heating systems

Where a gas-fired circulator is incorporated in the warm air unit to provide domestic hot water, it should be of a type that is able to deliver full- and part-load efficiency.	SBR-D-6.3.5
For a new gas-fired warm air system, the appliance should meet the recommendations of BS EN 778:1998 or BS EN 1319:1999, depending on the design of the appliance.	SBR-D-6.3.5
For ground to air and water to air systems, constant water flow should be maintained through the heat pump.	SBR-D-6.3.5

15.2.4 Hot water underfloor heating

The following controls should be fitted to ensure safe system operating temperatures:	SBR-D-6.3.8

- a separate flow temperature high limit thermostat for warm water systems connected to any high water temperature heat supply; and
- a separate means of reducing the water temperature to the underfloor heating system.

The minimum recommendations for room temperature, time and boiler controls are shown in Table 15.6.	SBR-D-6.3.8

Table 15.6 **Controls for underfloor heating**

Type of control	Means to achieve
Room temperature control	Thermostats for each room (adjacent rooms with similar functions may share) Weather compensating controller
Time control	Automatic setback of room temperature during unoccupied periods/at night
Boiler control	Boiler interlock

Bathrooms or en suites which share a heating circuit with an adjacent bedroom: SBR-D-6.3.8

- should provide heat only when the bedroom thermostat is activated;
- should be fitted with an independent towel rail or radiator.

15.2.5 Heat pumps

Heat pumps are at their most efficient when the source temperature is as high as possible, the heat distribution temperature is as low as possible and pressure losses are kept to a minimum.

If radiators are used they should be high-efficiency radiators with high water volume.   SBR-D-6.3.4

Supply water temperatures should be in the range 40°C to 55°C to radiators, 30°C to 40°C to an underfloor heating system and 35°C to 45°C to fan coil units.   SBR-D-6.3.4

Electrically driven heat pumps should have a coefficient of performance of not less than 2.0 when operating at the heating system design condition.   SBR-D-6.3.4

Heat pump controls

In new buildings, where space heating is provided by heating only (heat pumps or reverse-cycle heat pumps) the controls shown in Table 15.7 should be observed.   SBR-ND-6.3.5

For all systems in Table 15.7, additional controls should include room thermostats (if not integral heat pump) to regulate the space temperature and interlocked with the heat pump operation.   SBR-ND-6.3.5

Table 15.7 **Heat pump controls**

Source	System	Minimum controls package
All types	All technologies	On/off zone control. If the unit serves a single zone, and for buildings with a floor area of 150 m or less the minimum requirement is achieved by default time control
Air to air	Single package	Controls package for all types above plus heat pump unit controls to include: • control of room air temperature (if not provided externally); • control of outdoor fan operation; • defrost control of external airside heat exchanger; • control for secondary heating (if fitted)
Air to air	Split system, multi-split system, variable refrigerant flow system	Controls package for all types above plus heat pump unit controls to include: • control of room air temperature (if not provided externally); • control of outdoor fan operation; • defrost control of external airside heat exchanger; • control for secondary heating (if fitted)
Water or ground to air	Single package energy transfer systems (matching heating/cooling demand in buildings)	Controls package for all types above plus heat pump unit controls to include: • control of room air temperature (if not provided externally); • control of outdoor fan operation for cooling tower or dry cooler (energy transfer systems); • control for secondary heating (if fitted) on air to air systems; • control of external water pump operation
Air to water; water or ground to air	Single package, split package	Controls package for all types above plus heat pump unit controls to include: • control of water pump operation (if not provided externally); • control of outdoor fan operation for cooling tower or dry cooler (energy transfer systems); • control for secondary heating (if fitted); • control of external water pump operation
Gas engine-driven heat pumps	Multi-split. variable refrigerant flow	Controls package for all types above plus heat pump unit controls to include: • control of room air temperature (if not provided externally); • control of outdoor fan operation; • defrost control of external airside heat exchanger; • control for secondary heating (if fitted)

Controls for combined warm air and hot water systems

The following controls should be provided:	SBR-D-6.3.10

- independent time control of both the heating and hot water circuits (achieved by means of a cylinder thermostat and a timing device, wired such that when there is no demand for hot water both the pump and circulator are switched off);
- pumped primary circulation to the hot water cylinder;
- a hot water circulator interlock (achieved by means of a cylinder thermostat and a timing device, wired such that when there is no demand from the hot water both the pump and circulator are switched off); and
- time control by the use of either:
  - a full programmer with separate timing to each circuit;
  - two or more separate timers providing timing control to each circuit;
  - a programmable room thermostat(s) to the heating circuit(s); or
  - a time switch/programmer (two channel) and room thermostat.

15.2.6 Boilers

Oil storage tanks used solely to serve a fixed combustion appliance installation providing space heating facilities in a building shall ensure that they:	SBR-D-3.23, SBR-ND-3.23

- will inhibit fire from spreading to the tank and its contents from within, or beyond, the boundary;
- will reduce the risk of oil escaping from the installation; SBR-D-3.24, SBR-ND-3.24
- contain any oil spillage likely to contaminate any water supply, groundwater, watercourse, drain or sewer; and
- permit any spill to be disposed of safely.

Every building must be designed and constructed in such a way that the products of combustion are carried safely to the external air without harm to the health of any person through leakage, spillage or exhaust, nor permit the re-entry of dangerous gases from the combustion process of fuels into the building.

SBR-D-3.20, SBR-ND-3.20

Terminal discharge from condensing boilers

The condensate plume from a condensing boiler can cause damage to external surfaces of a building if the terminal location is not carefully considered. The manufacturer's instructions should be followed.

Solid fuel boilers

These should be thermostatically controlled to reduce the burning rate of the fuel, by varying the amount of combustion air to the fire.

SBR-D-6.3.8

For safety reasons, a suitable heat bleed (slumber circuit) from the system should be formed (e.g. a gravity-fed radiator without a TRV or a hot water cylinder that is connected independent of any controls).

SBR-D-6.3.8

For hot water systems, unless the cylinder is forming the slumber circuit, a thermostatically controlled valve should be fitted.

SBR-D-6.3.8

Boiler plant controls

When installing boiler plant in new buildings the minimum control package shown in Table 15.8 should be observed.

SBR-ND-6.3.4

Table 15.8 Minimum controls for new boilers or multiple-boiler systems (depending on boiler plant output or combined boiler plant output)

Boiler plant output and controls package	Minimum controls
<100 kW (package A)	Timing and temperature demand control which should be zone specific where the building floor area is greater than 150 m^2

(Continued)

Table 15.8 **Continued**

Boiler plant output and controls package	Minimum controls
	Weather compensation except where a constant temperature supply is required
100–500 kW (package B)	Controls package A above plus: Optimal start/stop control is required with night set-back or frost protection outside occupied periods Boiler with two-stage high/low firing facility or multiple boilers should be installed to provide efficient part-load performance For multiple boilers, sequence control should be provided and boilers, by design or application, should have limited heat loss from non-firing modules, for example by using isolation valves or dampers Individual boilers, by design or application, should have limited heat loss from non-firing modules, for example by using isolation valve or dampers
>500 kW (package C)	Controls package A and B above, plus the burner controls should be fully modulating for gas-fired boilers or multi-stage for oil-fired boilers

Electric boiler plant controls

See Table 15.9.

Table 15.9 **Electric boiler controls**

System	Controls
Boiler temperature control	The boiler should be fitted with a flow temperature control and be capable of modulating the power input to the primary water depending on space heating conditions Buildings with a total usable floor area up to 150 m^2 should be divided into at least two zones with independent temperature control For buildings with a total usable floor area greater than 150 m^2, sub-zoning of at least two space heating zones must be provided, each having separate timing and temperature controls, by either: • multiple heating zone programmers; or • a single multi-channel programmer
Zone temperature control	Separate temperature control of zones within the building, using either: • room thermostats or programmable room thermostats in all zones; • a room thermostat or programmable room thermostat in the main zone and individual radiators; • controls such as thermostatic radiator valves (TRVs) on all radiators in the other zones; or • a combination of the above

(Continued)

Table 15.9 Continued

System	Controls
Time control of space and water heating	Time control of space and water heating should be provided by either: • a full programmer with separate timing to each circuit; • two or more separate timers providing timing control to each circuit; or • programmable room thermostat(s) to the heating circuit(s), with separate timing of each circuit

15.2.7 Fuel storage

The guidance on oil relates only to its use solely where it serves a combustion appliance providing space heating (or cooking facilities) in a building. There is other legislation covering the storage of oils for other purposes. Heating oils comprise Class C2 oil (kerosene) or Class D oil (gas oil) as specified in BS 2869:2006.

Every building must be designed and constructed in such a way that an oil storage installation, incorporating oil storage tanks used solely to serve a fixed combustion appliance installation providing space heating (or cooking facilities) in a building, will inhibit fire from spreading to the tank and its contents from within, or beyond, the boundary.	SBR-D-3.23, SBR-ND-3.23

Oil storage

Every building must be designed and constructed in such a way that an oil storage installation, incorporating oil storage tanks used solely to serve a fixed combustion appliance installation providing space heating facilities in a building will: • reduce the risk of oil escaping from the installation; • contain any oil spillage likely to contaminate any water supply, groundwater, watercourse, drain or sewer; and • permit any spill to be disposed of safely.	SBR-D-3.24, SBR-ND-3.24

Liquefied petroleum gas storage

The operation of properties where liquefied petroleum gas (LPG) is stored or is in use is subject to legislation enforced by both the Health and Safety Executive (HSE) and the local authority.

Every building must be designed and constructed in such a way that each liquefied petroleum gas storage installation, used solely to serve a combustion appliance providing space heating, water heating (or cooking facilities) will: • be protected from fire spreading to any liquefied petroleum gas container; and • not permit the contents of any such container to form explosive gas pockets in the vicinity of any container.	SBR-D-4.11, SBR-ND-4.11

Woody biomass storage

By its very nature woody biomass fuel is highly combustible and precautions need to be taken to reduce the possibility of the stored fuel igniting.

To maintain fireproof storage and prevent back-burning there should be an interruption to the fuel transport system, normally by use of a star-feeder or chute for the fuel to fall into the boiler (see BS EN 303-5:1999).	SBR-D-3.23.4, SBR-ND-3.23.4
The woody biomass fuel should be stored separately from the boiler that the fuel feeds for fire safety reasons.	SBR-D-3.23.4, SBR-ND-3.23.4

15.2.8 Work on existing buildings

Where a new or replacement boiler storage vessel is installed, or where existing systems are extended, new or existing pipes that are accessible or exposed as part of the work should be insulated as for new systems.	SBR-D-6.4.3, SBR-ND-6.4.3
Where alterations are being made to an existing heating system or a new or replacement heating system is being installed in an existing dwelling	SBR-D-6.3.12

(or building consisting of dwellings), such alterations
should not allow the heating system as a whole to
be downgraded in terms of energy efficiency or
compromised from a safety point of view.

Additional guidance for hospitals

Boiler houses should: SBR-ND-2.B.1

- never be directly below, nor directly adjoin,
  the operating theatres, intensive therapy units
  or special care baby units; and
- be provided with a fire suppression system
  if they are directly below, or directly adjoin,
  any other hospital department to which
  patients have access.

Additional guidance for enclosed shopping centres

On the operation of the fire alarm (subject to the   SBR-ND-2.C.5
4-minute grace period where appropriate) all
air moving systems, mains and pilot gas outlets,
combustion air blowers and gas, electrical and
other heating appliances in the reservoir shall be
shut down.

15.2.9 Efficiency and credits

Appliances installed in a building should be energy   SBR-ND-6.3.1
efficient.

Designers of non-domestic buildings may wish to consider using heating
efficiency credits when designing systems incorporating boilers, warm air heaters, radiant heaters and heat pumps to exceed the minimum efficiency specified.

Examples of how this is achieved are given in Annex 6F to the Scottish
Building Regulations.

Appliance efficiency

The following tables recommend efficiencies for:

- minimum boiler seasonal efficiency for heating plant (Tables 15.10 and 15.11);
- minimum thermal efficiency for gas- and oil-fired warm air systems and radiant heaters (Tables 15.12 and 15.13); and
- coefficient of performance (COP) for heat pumps (Table 15.14).

Table 15.10 **Boiler seasonal efficiency in new buildings**

Fuel type	Boiler system	Minimum boiler seasonal efficiency[a]
Gas (natural)	Single	84%
Gas (LPG)	Multiple	80% for any individual boiler and 84% for the overall multi-boiler system

[a]Based on gross calorific value.

Table 15.11 **Minimum boiler seasonal efficiency**

Fuel type	Minimum effective heat-generating seasonal efficiency[a]	Effective heat-generating seasonal efficiencies and boiler seasonal efficiency in existing buildings[a]
Gas (natural)	84%	80%
Gas (LPG)	85%	81%
Oil	86%	82%

[a]Based on gross calorific value.

Table 15.12 **Gas and oil-firing warm air systems: minimum thermal efficiency**

System	Minimum thermal efficiency[a]
Gas-firing forced convection heater without a fan complying with EN 621	80%
Fan assisted gas-firing forced convection complying with EN 1020	80%
Direct gas-firing forced convection heater complying with EN 525	90%
Oil-firing forced convection	80%

[a]Based on gross calorific value.

Table 15.13 **Radiant heaters: minimum thermal efficiency**

System	Minimum thermal efficiency[a]
Luminous (flueless)	85.5%
Non-luminous (flueless)	85.5%
Non-luminous (flued)	73.8%
Multi-burner radiant heaters	80%

[a]Based on gross calorific value.

Table 15.14 **Heat pump coefficient of performance (COP)**

System	Minimum heating COP (at design condition)
All types except absorption heat pumps and gas engine heat pumps	2.0
Absorption heat pumps	0.5
Gas engine-driven heat pumps	1.0

Combined heat and power quality index

The Combined Heat and Power Quality Assurance (CHPQA) programme is a registration and certification scheme which serves as an indicator of the energy efficiency and environmental performance of a combined heat and power (CHP) scheme, relative to the generation of the same amounts of heat and power by separate, alternative means.

The required minimum combined heat and power quality index for all types of CHP should be 105.	SBR-ND-6.3.3

There is no minimum combined heat and power quality index specified for electric (primary) heating.

The CHP unit should operate as the lead heat generator and be sized to supply no less than 45% of the annual heating demand.	SBR-ND-6.3.3
CHP may be used as the main or supplementary heat source in community heating or district heating schemes.	SBR-ND-6.3.3

Space heating controls (general)

> If space heating is to be intermittent (and does   SBR-D-6.3.8
> not make use of off-peak electricity) the system
> should only operate when the building is normally
> occupied or is about to be occupied.

15.2.10 Commissioning building services

Commissioning (i.e. in terms of achieving the levels of energy efficiency that the component manufacturers expect from their product(s)) should also be carried out with a view to ensuring the safe operation of the system.

> Every building must be designed and constructed in   SBR-D-6.7,
> such a way that energy supply systems and building   SBR-ND-6.7
> services which use fuel or power for heating the
> internal environment are commissioned to achieve
> optimum energy efficiency.

15.2.11 Written information

> For a domestic building, written information   SBR-D-6.8.1
> concerning the operation and maintenance of the
> heating system (together with any decentralized
> power generation equipment) should be made
> available for the use of the occupier.
> For non-domestic buildings, a logbook containing   SBR-ND-6.8.1
> information on energy system operation and
> maintenance (such as building services plant
> and controls) shall be provided to ensure that the
> building user can optimize the use of fuel.

The Chartered Institution of Building Services Engineers (CIBSE) Technical Memorandum 31 (TM31) provides guidance on the presentation of a logbook, and the logbook information should be presented in this or a similar manner.

Where a building contains multiple dwellings a rating   SBR-D-6.9.1,
is required for each individual dwelling, taking into   SBR-ND-6.9.1
consideration the percentage of low-energy lighting
and the type of heating that has been installed.

15.2.12 Metering

To enable building operators to manage fuel use effectively, systems should
be provided with fuel meters to enable the annual fuel consumption to be
accurately measured.

Where multiple buildings or fire-separated units are   SBR-ND-6.10.1
served on a site by a communal heating appliance,
the fuel metering shall be installed at both the
communal heating appliance and heat meters at the
individual buildings served.
A fuel meter should be installed if a new fuel type   SBR-ND-6.10.2
or new boiler (where none existed previously) is
installed.

15.2.13 Conservatories

As a conservatory which is heated will be inefficient in energy terms, the
general guidance to occupiers is that conservatories should be heated as little
as possible.

Conservatories with heating installed should have   SBR-D-6.3.13
controls (such as a TRV to the radiator) to regulate
the heating separately from the rest of the dwelling.

Annex A
Acronyms and Abbreviations

ANSI	American National Standards Institute
BEAB	British Electrotechnical Approvals Board
BMS	Building Management System
BR	Building Regulations
BRE	Building Research Establishment
BS	British Standard
BSI	British Standards Institution
CE	Conformité Européene
CEN	Comité Européen de Normalization
CENELEC	Comité Européen de Normalization Electrotechnique
CHP	Combined Heat and Power
CHPQA	Combined Heat and Power Quality Assurance
CIBSE	Chartered Institution of Building Services Engineers
CITB	Construction Industry Training Board
CO	Carbon Monoxide
CO_2	Carbon Dioxide
COP	Coefficient Of Performance
CORGI	Council for Registered Gas Installers
CPSU	Combined Primary Storage Unit
D	domestic
DDA	Disability Discrimination Act 2005
DHW	Domestic Hot Water
DIY	Do It Yourself
DPM	Damp-Proof Membrane
DWD	Drinking Water Directive
DWI	Drinking Water Inspectorate
EC	European Community
EEC	European Economic Commission
EMC	Electromagnetic Compatibility
EN	European Normalization
ENV	European Pre-standard

EU	European Union
GoCo	Government-owned Company
GRP	Glass Reinforced Plastic
H_2	Hydrogen
H_2O	Water
HETAS	Heating Equipment Testing Approval Scheme
HSE	Health & Safety Executive
ISO	International Standards Organization
LPG	Liquefied Petroleum Gas
LPGA	Liquefied Petroleum Gas Association
LZCT	Low and Zero Carbon Technologies
ND	Non-domestic
NFS	The Public Health and Safety Company
NSF	National Sanitation Foundation
NSO	National Standards Organization
prEN	European Draft Standards
PSV	Passive Stack Ventilation
RPZ	Reduced Pressure Zone
SAI	System Assurance Institute
SBR	Scottish Building Regulations
SEPA	Scottish Environmental Protection Agency
SI	Statutory Instrument
SNIPEF	Scottish and Northern Ireland Plumbing Employers Federation
SUDS	Sustainable Urban Drainage Systems
TRV	Thermostatic Radiator Valve
UKAS	United Kingdom Accreditation Service
uPVC	unplasticized PolyVinyl Chloride
WC	Water Closet
WRAS	Water Regulations Advisory Scheme
WRc plc	Water Research Centre

Annex B
Legislation and Standards

EU Harmonized Directives

CE Marking Directive (93/68/EEC)
Construction Products Directive (89/106/EEC) (SI 1994/3051)
Drinking Water Directive (DWD 98/83/EC)
EC Dangerous Substances Directive (76/464/EEC)
EMC Directive (89/336/EEC)
European Water Framework Directive (2006/118/EC)
General Rules and Rules for Buildings in Concrete and Steel DD ENV 1992-1-1:
1992 Eurocode 2: Part 1 and DD ENV 1993-1-1:1992 Eurocode 3: Part 1-1
Low Voltage Directive (73/23/EEC and amendment 93/68 EEC)
Workplace (Health, Safety and Welfare) Regulations 1992

UK Legislation

Building (Scotland) Act 2003
Building (Scotland) Regulations 2004
Building Act 1984
Building Regulations (Northern Ireland) 2000
Disability Discrimination Act 1995
Factories Act 1961
Food Hygiene (General) Regulations 1970
Gas Safety (Installations and Use) Regulations 1998
Northern Ireland Water Regulations
Offices, Shops and Railway Premises Act 1963
Planning and Building Regulations (Amendment) (NI) Order 1990)
Public Health Act 1936
Requirements for Water Fittings, of the Water Supply (Water Fittings) Regulations 1999
School Premises (General Requirements and Standards) (Scotland) Regulations 1967
Scottish and Northern Ireland Plumbing Employers' Federation (SNIPEF)

Scottish Water Bye-laws 2004
Sewerage (Scotland) Act 1968
Water Act 1945
Water (Scotland) Act 1980
Water Bye-laws 2000 (Scotland)
Water Fittings and Materials Directory
Water Industry Act 1991
Water Supply (Water Fittings) Regulations 1999
Workplace (Health, Safety and Welfare) Regulations 1992

Standards

Anti-flooding valves	PREn 13564
Cast iron pipes and fittings, their joints and accessories for the evacuation of water from buildings Requirements, test methods and quality assurance	BS EN 877:1999
Code of practice for building drainage	BS 8301:1985
Code of practice for design and installation of small sewage treatment works and cesspools	BS 6297:1983
Code of practice for drainage of roofs and paved areas	BS 6367:1983
Code of practice for protection of structures against water from the ground	BS 8102:1990
Code of practice for sanitary pipework	BS 5572:1991
Construction and testing of drains and sewers	BS EN 1610:1998
Copper and copper alloys. Plumbing fittings	BS EN 1254:1998
Part 1: Fittings with ends for capillary soldering or capillary brazing to copper tubes	
Part 2: Fittings with compression ends for use with copper tubes	
Part 3: Fittings with compression ends for use with plastic pipes	
Part 4: Fittings combining other end connections with capillary or compression ends	
Part 5: Fittings with short ends for capillary brazing to copper tubes	
Copper and copper alloys. Seamless, round copper tubes for water and gas in sanitary and heating applications	BS EN 1057:1996
Copper indirect cylinders for domestic purposes	BS 1566-1:1984
Drain and sewer systems outside buildings	BS EN 752

Part 1: 1996 Generalities and definitions
Part 2: 1997 Performance requirements
Part 3: 1997 Planning
Part 4: 1997 Hydraulic design and environmental aspects
Part 5: 1997 Rehabilitation
Part 6: 1998 Pumping installations
Part 7: 1998 Maintenance and operations

Ductile iron pipes, fittings, accessories and their joints for sewerage applications. Requirements and test methods	BS EN 598:1995
Gravity drainage systems inside buildings	BS EN 12056:2000

Part 1: Scope, definitions, general and performance requirements
Part 2: Wastewater systems, layout and calculation
Part 3: Roof drainage layout and calculation
Part 4: Effluent lifting plants, layout and calculation
Part 5: Installation, maintenance and user instructions

Heating boilers. Heating boilers with forced draught burners. Terminology, general requirements, testing and marketing	BS EN 303-1:1999
Plastics piping systems for soil and waste discharge (low and high temperature) within the building structure	BS EN 1566-1:2000
Protection of buildings against water from the ground	BS CP 102:1973
Small wastewater treatment plants less than 50 PE	BS EN 12566-1:2000
Specification for copper hot water storage combination units for domestic purposes	BS 3198:1981
Specification for flexible joints for grey or ductile cast iron drain pipes and fittings (BS 437) and for discharge and ventilating pipes and fittings (BS 416)	BS 6087:1990
Specification for installation of hot water supplies for domestic purposes, using gas fired appliances of rated input not exceeding 70 kW	BS 5546:2000
Vitreous china washdown WC pans with horizontal outlet. Specification for WC pans with horizontal outlet for use with 7.5 maximum flush capacity cisterns	BS 5503-3:1990

| Wall hung WC pan, specification for wait hung WC pans with horizontal outlet for use with 7.5 maximum flush capacity cisterns | BS 5504-4:1990 |

Copies of all British Standards are available from: BSI, PO Box 16206, Chiswick, London W4 4ZL. Website: www.bsonline.techindex.co.uk

Index